MEASURING WHAT COUNTS

A CONCEPTUAL GUIDE FOR MATHEMATICS ASSESSMENT

CARL CAMPBELL BRIGHAM LIBRARY
EDUCATIONAL TESTING SERVICE
PRINCETON, NJ 08541

B01V

MATHEMATICAL SCIENCES
EDUCATION BOARD
NATIONAL RESEARCH COUNCIL

NATIONAL ACADEMY PRESS   WASHINGTON, DC 1993

NATIONAL ACADEMY PRESS
2101 Constitution Avenue, NW • Washington, DC 20418

NOTICE: The project that is the subject of this report was approved by the Governing Board of the National Research Council, whose members are drawn from the councils of the National Academy of Sciences, the National Academy of Engineering, and the Institute of Medicine. The members of the committee responsible for the report were chosen for their special competences and with regard for appropriate balance.

This report has been reviewed by a group other than the authors according to procedures approved by a Report Review Committee consisting of members of the National Academy of Sciences, the National Academy of Engineering, and the Institute of Medicine.

The National Research Council was organized by the National Academy of Sciences in 1916 to associate the broad community of science and technology with the Academy's purposes of furthering knowledge and advising the federal government. Functioning in accordance with general policies determined by the Academy, the Council has become the principal operating agency of both the National Academy of Sciences and the National Academy of Engineering in providing services to the government, the public, and the scientific and engineering communities. The Council is administered jointly by both Academies and the Institute of Medicine. Dr. Bruce M. Alberts and Dr. Robert M. White are chairman and vice chairman, respectively, of the National Research Council.

The Mathematical Sciences Education Board was established in 1985 to provide a continuing national capability to assess the status and quality of education in the mathematical sciences and is concerned with excellence in education for all students at all levels. The Board reports directly to the Governing Board of the National Research Council.

Library of Congress Catalog Card Number 93-85917
International Standard Book Number 0-309-04981-4

Copyright 1993 by the National Academy of Sciences. All rights reserved.

Permission for limited reproduction of portions of this book for educational purposes but not for sale may be granted upon receipt of a written request to the National Academy Press, 2101 Constitution Avenue, NW, Washington, DC 20418.

Copies of the report may be purchased from the National Academy Press, 2101 Constitution Avenue, NW, Box 285, Washington, DC 20055. Call 800-624-6242 or 202-334-3313 (in the Washington Metropolitan Area).

Printed in the United States of America

B122

CONTENTS

PREFACE

Calls for standards in education have been echoing across the nation for several years, especially since political leaders of both parties decided to adopt bipartisan national goals for education. Standards without appropriate means of measuring progress, however, amount to little more than empty rhetoric. To stay the course and achieve the national goals for education, we must measure the things that really count.

Standards take many forms and appear under many guises. Curriculum (or content) standards tell what students should learn. Teaching (or pedagogical) standards tell how students learn and how teachers should teach. Delivery (or opportunity-to-learn) standards tell what is necessary of schools so that students can learn and teachers can teach. Assessment (or performance) standards tell what students should know and be able to do as well as how evaluators can judge levels of performance.

Since 1989 mathematics has led the national movement towards standards with *Everybody Counts* (National Research Council, 1989), *Curriculum and Evaluation Standards for School Mathematics* (National Council of Teachers of Mathematics (NCTM), 1989), and *Professional Standards for Teaching Mathematics* (NCTM, 1991). In April 1991, the Mathematical Sciences Education Board (MSEB) convened a national summit on assessment, which led to *For Good Measure* (1991), a concise statement of goals and objectives for mathematics assessment. To move the national discussion from generalities to specifics, the MSEB then published *Measuring Up* (1993), which provided prototype assessment tasks for fourth grade mathematics that illustrate in concrete terms the goals of the NCTM *Standards*.

Measuring What Counts further advances this national discussion by establishing crucial research-based connections between standards and assessment. It demonstrates the importance of three key principles—on content, learning, and equity—for any program of assessment that is intended to support the national educational goals. The message of *Measuring What Counts* is quite simple, but its implications are profound: Assessment in support of standards must not only measure results, but must also contribute to the educational process itself.

The analyses and recommendations in this report were developed by a study group formed by the NRC in 1991 to develop conceptual guidelines that would make content the driving force in the reform of mathematics assessment and that would explore a variety of related measurement and policy issues. The intent of *Measuring What Counts* is not to offer immediate practical advice, but to lay out a conceptual framework that will help those who are struggling with the urgent need to develop new assessments that align properly with the national standards for mathematics education.

The three principles on content, learning, and equity articulated in *Measuring What Counts* are necessary but not sufficient criteria for effective assessment. They set forth fundamental conditions that form a foundation on which to build new approaches to traditional technical testing issues such as reliability and validity. Assessment should foremost reflect important mathematics, support good instructional practice, and enhance every student's opportunity to learn.

Although these principles are rooted in both informed practice and extensive research, it is fair to say that there remain many open questions. Research shows clearly that the task of assessing mathematical learning is far more subtle than previously believed; experience reveals enormous gaps between current assessment practice and new goals for mathematics education. It is clear from the recent history of failed reform that when assessment is separated from curriculum and instruction, teaching becomes distorted, thus diminishing learning.

Experts agree that for education to be effective, curriculum, instruction, and assessment must harmonize for their mutual support. Both internal (teacher-based) and external (district- or state-based) assessment must support improved learning. However, the path from general agreement to specific assessments is far from clear. We are embarking on a new venture, guided by the principles of content, learning, and equity. Exploration of this new world of alternative assessment will take years of work from thousands of practitioners working with mathematics education specialists and measurement experts to achieve a more effective balance of assessment in practice.

Although many of the issues raised in this report apply to all disciplines, it is mathematics education that provides the primary motivation for the study, the background of the authors, the source of examples, the domain of research, and the field of practice on which the conclusions and recommendations are based. *Measuring What Counts* seeks to address issues in assessment that are important to the discipline of mathematics and about which the expertise of mathematics educators can make a singular contribution. Content, learning, and equity emerged as fundamental principles for assessment because they are fundamental concerns of mathematics education.

One consequence of *Measuring What Counts* should be a new wave of research on assessment, on learning, and on instruction. Since much that is in this report is based on expert conjecture rather than firm evidence, it opens scores of potential areas for further research. Indeed, the changing practice of mathematics itself—the increased focus on computer-enhanced work, on group problem solving, on modeling complex problems—challenges researchers in assessment and learning with issues rarely before considered. The resulting iteration of practice and research will provide an effective guidance system to keep assessment reform aligned with curricular objectives and principles of learning.

All reform is evolutionary. As society changes, the targets and goals for education change. Assessment is our primary tool for monitoring progress and making midcourse corrections. The principles of assessment set forth in *Measuring What Counts* provide a solid conceptual basis for current efforts to improve assessment and lay the groundwork for more detailed assessment standards to be published by the NCTM.

When the stakes for improved education are so high, when our children's futures are at stake, we must ensure that assessment supports standards-based education by adhering to fundamental principles of content, learning, and equity.

ACKNOWLEDGMENTS

Measuring What Counts was prepared by the National Research Council's Study Group on Mathematics Assessment, which was chaired by Jeremy Kilpatrick, Regents Professor of Mathematics Education at the University of Georgia. The group met over a period of two years to develop drafts of this report. Members of the study group, in addition to the chair, were Janice Arceneaux, Magnet Specialist in the Houston Independent School District; Lloyd Bond, Professor of Educational Research Methodology at the University of North Carolina-Greensboro; Felix Browder, Professor of Mathematics at Rutgers University; Philip C. Curtis, Jr., Professor Mathematics at the University of California at Los Angeles; Jane D. Gawronski, Superintendent of the Escondido Union High School District; Robert L. Linn, Professor of Education at the University of Colorado-Boulder; Sue Ann McGraw, Mathematics Teacher at Lake Oswego High School; Robert J. Mislevy, Principal Research Scientist at Educational Testing Service; Alice Morgan-Brown, Statewide Director for Academic Champions of Excellence Program at Morgan State University; Andrew Porter, Director, Wisconsin Center for Education Research at the University of Wisconsin-Madison; Marilyn Rindfuss, National Mathematics Consultant at The Psychological Corporation; Edward Roeber, Director, Student Assessment Programs at the Council of Chief State School Officers; Maria Santos, Mathematics and Science Supervisor in the San Francisco Unified School District; Cathy Seeley, Director of Pre-college Programs, Charles A. Dana Center for Mathematics and Science Education at the University of Texas-Austin; and Edward A. Silver, Professor and Senior Scientist, University of Pittsburgh.

Members of the study group faced many significant hurdles posed by their differing professional perspectives, by the rapidly changing context of educational assessment, and by the challenges posed by the new *Standards* for school mathematics. We owe each of them a special thanks for persisting in this formidable task to reach consensus on the key principles enunciated in this report. Particular thanks are due Jeremy Kilpatrick not only for his able leadership as chairman of the Study Group, but also for the substantial contribution he made in writing and editing the various parts of the report.

The work of the study group was enriched by three commissioned papers that appear at the end of this volume. These papers provided useful background for deliberations of the study group and constituted valuable additions to the research literature on assessment. We are particularly grateful to their authors—Lynn Hancock and Jeremy Kilpatrick, Stephen B. Dunbar and Elizabeth A. Witt, and Diana C. Pullin—for their contributions to this endeavor. We wish also to thank Linda Dager Wilson and Lynn Hancock for supplying other background information of value to this study.

Like all reports of the National Research Council, *Measuring What Counts* has been extensively reviewed—first by outside experts in early draft form, then by the MSEB Committee on Policy Studies at several key stages, and, at the final stage, under the careful protocol of the NRC's Report Review Committee. We thank these many reviewers for their insightful and knowledgeable comments. Special thanks are due Nancy Cole, Chair of the MSEB Committee on Policy Studies, for providing consistent and wise counsel as the report worked its way through various drafts. *Measuring What Counts* is much stronger as a result of the input and advice of these outside reviewers.

Financial support for work of the study group and preparation of *Measuring What Counts* was provided by the U.S. Department of Education and the National Science Foundation. We gratefully acknowledge the support of these organizations.

Staff work in support of the study group was provided by the MSEB Office of Policy Studies directed by Linda Peller Rosen, Associate Executive Director of the MSEB. Edward T. Esty and Patricia A. Butler deserve special thanks for managing this complex project with unfailing perseverance and tireless energy. Thanks are due also to Anuradha Sapru Kohls, Ramona Robertson, and Altoria Bell, who undertook diverse tasks without which this report would never have been completed. The staff of the National Academy Press deserves special mention for their efforts on our behalf.

Hyman Bass

Hyman Bass, Chairman
Mathematical Sciences Education Board

MEASURING WHAT COUNTS

EXECUTIVE SUMMARY

"You can't fatten a hog by weighing it." So said a farmer to a governor at a public hearing in order to explain in plain language the dilemma of educational assessment. To be useful to society, assessment must advance education, not merely record its status.

Assessment is a way of measuring what students know and of expressing what students should learn. As the role of mathematics in society has changed, so mathematics education is changing, based on new national standards for curriculum and instruction. Mathematics assessment must also change to ensure consistency with the goals of education.

Three fundamental educational principles form the foundation of all assessment that supports effective education:

THE CONTENT PRINCIPLE
Assessment should reflect the mathematics that is most important for students to learn.

THE LEARNING PRINCIPLE
Assessment should enhance mathematics learning and support good instructional practice.

THE EQUITY PRINCIPLE
Assessment should support every student's opportunity to learn important mathematics.

Despite their benign appearance, these principles contain the seeds of revolution. Few assessments given to students in America today reflect any of these vital principles. For educational

Assessment is a way of measuring what students know and of expressing what students should learn.

reform to succeed, the yardsticks of progress must be rooted in the principles of content, learning, and equity.

MATHEMATICS IN TODAY'S WORLD

The pressures to change mathematics education reflect society's disappointment with the lack of interest and accomplishment of so many students in today's schools. In the background of public debate is the steady criticism that school mathematics is out of step with today's world and is neither well taught nor well learned.

Unfortunately, these pressures often suggest inconsistent courses of action, with standards-based curriculum and instruction moving in one direction while mandated tests remain aimed in another direction, at an older, more traditional target. Too often, teachers are caught in the middle. To be effective, mathematics education must be rooted in the practice of mathematics, in the art of teaching, and in the needs of society. These pivotal forces drive the current movement to improve mathematics education:

- A more comprehensive view of mathematics and its role in society: mathematics is no longer just a prerequisite subject for prospective scientists and engineers but is a fundamental aspect of literacy for the twenty-first century.

- A recommitment to the traditional wisdom that mathematics must be made meaningful to students if it is to be learned, retained, and used.

- The growing recognition that in this technological era, all students should learn more and better mathematics.

ASSESSMENT IN TODAY'S WORLD

Assessment is the guidance system of education just as standards are the guidance system of reform. It helps teachers and parents determine what students know and what they need to learn. Assessment can play a powerful role in conveying clearly and directly how well students are learning and how well school systems are responding to the national call for higher education standards.

> Assessment is the guidance system of education just as standards are the guidance system of reform.

At its root, assessment is a communication process that tells students, teachers, parents, and policymakers some things—but not everything—about what students have learned. Assessment provides information that can be used to award grades, to evaluate a curriculum, or to decide whether to review fractions. Internal assessment communicates to teachers critical aspects of their students' performance, helping them to adjust their instructional techniques accordingly. External assessment provides information about mathematics programs to parents, state and local education agencies, funding bodies, and policymakers.

Many reformers see assessment as much more than an educational report card. Assessment can be the engine that propels reform forward, but only if we make *education* rather than *measurement* the driving force in the development of new assessments. By setting a public and highly visible target to which all can aspire, assessment can inform students, parents, and teachers about the real performance-based meaning of curriculum guidelines. Assessments not only measure what students know but provide concrete illustrations of the important goals to which students and teachers can aspire.

Assessment can

be the engine

that propels

reform forward,

but only if

education

rather than

measurement is

the driving force.

ASSESSMENT IN THE SERVICE OF EDUCATION

Improved assessment is required to complement and support the changes under way in mathematics education: both in the kinds of mathematics that are taught and in the ways in which they are taught. As such, assessment is an integral part of an interlocking triad of reforms along with curriculum and professional development of teachers. Because assessment is key to determining what students learn and how teachers teach, it must be reshaped in a manner consistent with the new vision of teaching and learning.

Students learn important mathematics when they use mathematics in relevant contexts in ways that require them to apply what they know and extend their thinking. Students think when they are learning and they learn when they are thinking. Good teachers have long recognized that mathematics comes alive for students when it is learned through experiences they find meaningful and valuable. Students learn best and most enduringly by engaging mathematics actively, by reflecting on their experience, and by communicating with others about it. Students want to make sense of the world, and mathematics is a wonderful tool to use in this eternal quest.

Because teamwork is important on the job and in the home, mathematics students learn important lessons when they work in teams, combining their knowledge and discovering new ways of solving problems. Often there is no single right answer, only several possibilities that unfold into new questions. Students need opportunities to advance hypotheses, to construct mathematical models, and to test their inferences by using the mathematics of estimation and uncertainty alongside more traditional techniques of school mathematics. Hand-held graphing calculators allow, for the first time, thorough exploration of complex, real-life problems. Computational impediments need no longer block the development of problem-solving or mathematical modeling skills.

This new vision of learning and teaching is now being tried in some classrooms across the country. Current assessment does not support this vision and often works against it. For decades, educational assessment in the United States has been driven largely by practical and technical concerns rather than by educational priorities. Testing as we know it today arose because very efficient methods were found for assessing large numbers of people at low cost. A premium was placed on assessments that were easily administered and that made frugal use of resources. The constraints of efficiency meant that mathematics assessment tasks could not tap a student's ability to estimate the answer to an arithmetic calculation, construct a geometric figure, use a calculator or ruler, or produce a complex deductive argument.

A narrow focus on technical criteria—primarily reliability— also worked against good assessment. For too long, reliability meant that examinations composed of a small number of complex problems were devalued in favor of tests made up of many short items. Students were asked to perform large numbers of smaller tasks, each eliciting information on one facet of their understanding, rather than to engage in complex problem solving or modeling, the mathematics that is most important.

In the absence of expressly articulated educational principles to guide assessment, technical and practical criteria have become de facto

For decades, educational assessment in the United States has been driven largely by practical and technical concerns rather than by educational priorities.

ruling principles. The content, learning, and equity principles are proposed not to challenge the importance of these criteria, but to challenge their dominance and to strike a better balance between educational and measurement concerns. An increased emphasis on validity—with its attention to fidelity between assessments, high-quality curriculum and instruction, and consequences—is the tool by which the necessary balance can be achieved.

In some ways, test developers do acknowledge the importance of curricular and educational issues. However, their concern is usually about coverage, so they design tests by following check-off lists of mathematical topics (e.g., fractions, single-digit multiplication). This way of determining test content matched fairly well the old vision of mathematics instruction. In this view you could look at little pieces of learning, add them up, and get the big picture of how well someone knew mathematics.

Today we recognize that students must learn to reason, create models, prove theorems, and argue points of view. Assessments must reflect this recognition by adhering to the three principles of content, learning and equity. You cannot get at this kind of deep understanding and use of mathematics by examining little pieces of learning. Assessments that are appropriately rich in breadth and depth provide opportunities for students to demonstrate their deep mathematical understanding. Mathematics education and mathematics assessment must be guided by a common vision.

THE CONTENT PRINCIPLE

Any assessment of mathematics learning should first and foremost be anchored in important mathematics. Assessment should do much more than test discrete procedural skills so typical of today's topic-by-process frameworks for formal assessments. Many current assessments distort mathematical reality by presenting mathematics as a set of isolated, disconnected fragments, facts, and procedures. The goal ought to be assessment tasks that elicit student work on the meaning, process, and uses of mathematics.

Important mathematics must shape and define the content of assessment. Appropriate tasks emphasize connections within mathematics, embed mathematics in relevant external contexts, require students to communicate clearly their mathematical thinking,

Assessment should reflect the mathematics that is most important for students to learn.

and promote facility in solving nonroutine problems. Considerations of connections, communication, and nonroutine problems raise many thorny issues that testmakers and teachers are only beginning to explore. However, these considerations are essential if students are to meet the new expectations of mathematics education standards.

The content principle has profound implications for those who design, score, and use mathematics assessments. Many of the assessments used today, such as standardized multiple-choice tests, have reinforced the view that the mathematics curriculum should be constructed from lists of narrow, isolated skills that can be easily disassembled for appraisal. The new vision of school mathematics requires a curriculum and matching assessment that is both broader and more integrated.

The mathematics in an assessment must never be distorted or trivialized for the convenience of assessment. Assessment should emphasize problem solving, thinking, and reasoning. In assessment as in curriculum activities, students should build models that connect mathematics to complex, real-world situations and regularly formulate problems on their own, not just solve those structured by others. Rather than forcing mathematics to fit assessment, assessment must be tailored to the mathematics that is important to learn.

Implications of the content principle extend as well to the scoring and reporting of assessments. New assessments will require new kinds of scoring guides and ways of reporting student performance that more accurately reflect the richness and diversity of mathematical learning than do the typical single-number scores of today.

THE LEARNING PRINCIPLE

To be effective as part of the educational process, assessment should be seen as an integral part of learning and teaching rather than as the culmination of the process. Time spent on assessment will then contribute to the goal of improving the mathematics learning of all students.

If assessment is going to support learning, then assessment tasks must provide genuine opportunities for all students to learn significant mathematics. Too often a sharp line has been drawn

Rather than forcing mathematics to fit assessment, assessment must be tailored to the mathematics that is important to learn.

Assessment should enhance mathematics learning and support good instructional practice.

between assessment and instruction. Teachers teach, then instruction stops and assessment occurs. In the past, for example, students' learning was often viewed as a passive process whereby students remember what teachers tell them to remember. Consistent with this view, assessment has often been thought of as the end of learning. The student is assessed on material learned previously to see if her or she remembers it. Earlier conceptions of the mathematics curriculum as a collection of fragmented knowledge led to assessment that reinforced the use of memorization as a principal learning strategy.

Today we recognize that students make their own mathematics learning individually meaningful. Learning is a process of continually restructuring prior knowledge, not just adding to it. Good education provides opportunities for students to connect what is being learned to prior knowledge. Students know mathematics if they have developed the structures and meanings of the content for themselves.

If assessment is going to support good instructional practice, then assessment and instruction must be better integrated than is commonly the case today. Assessment must enable students to construct new knowledge from what they know. The best way to provide opportunities for the construction of mathematical knowledge is through assessment tasks that resemble learning tasks in that they promote strategies such as analyzing data, drawing contrasts, and making connections. This can be done, for example, by basing assessment on a portfolio of work that the student has done as part of the regular instructional program, by integrating the use of scoring guides into instruction so that students will begin to internalize the standards against which the work will be evaluated, or by using two-stage testing in which students have an extended opportunity to revise their initial responses to an assessment task.

Not only should all students learn some mathematics from assessment tasks, but the results should yield information that can be used to improve students' access to subsequent mathematical knowledge. The results must be timely and clearly communicated to students, teachers, and parents. School time is precious. When students are not informed of their errors and misconceptions, let alone helped to correct them, the assessment may both reinforce misunderstandings and waste valuable instructional time.

> Assessment
> tasks must
> provide genuine
> opportunities for
> all students to
> learn significant
> mathematics.

When the line between assessment and instruction is blurred, students can engage in mathematical tasks that not only are meaningful and contribute to learning, but also yield information the student, the teacher, and perhaps others can use. In fact, an oft-stated goal of reform efforts in mathematics education is that visitors to classrooms will be unable to distinguish instructional activities from assessment activities.

THE EQUITY PRINCIPLE

The idea that some students can learn mathematics and others cannot must end; mathematics is not reserved for the talented few, but is required of all to live and work in the twenty-first century. Assessment should be used to determine what students have learned and what they still need to learn to use mathematics well. It should not be used to filter students out of educational opportunity.

Designing assessments to enhance equity will require conscientious rethinking not just of what we assess and how we do it but also of how different individuals and groups are affected by assessment design and procedures. The challenge posed by the equity principle is to devise tasks with sufficient flexibility to give students a sense of accomplishment, to challenge the upper reaches of every student's mathematical understanding, and to provide a window on each student's mathematical thinking.

Some design strategies are critical to meeting this challenge, particularly permitting students multiple entry and exit points in assessment tasks and allowing students to respond in ways that reflect different levels of mathematics knowledge or sophistication. But there are no guarantees that new assessment will be fairer to every student, that every student will perform better on new assessments, or that differences between ethnic, linguistic, and socioeconomic groups will disappear. While this is the hope of the educational reform community, it seems clear that hope must be balanced by a spirit of empiricism: there is much more to be learned about how changes in assessment will affect longstanding group differences.

Equity implies that every student must have an opportunity to learn the important mathematics that is assessed. Obviously, students who have experience reflecting on the mathematics they are learning, presenting and defending their ideas, or organizing,

Assessment should support every student's opportunity to learn important mathematics.

executing, and reporting on a complex piece of work will have an advantage when called upon to do so in an assessment situation. Especially when assessments are used to make high-stakes decisions on matters such as graduation and promotion, the equity principle requires that students be guaranteed certain basic safeguards. Students cannot be assessed fairly on mathematics content that they have not had an opportunity to learn.

Assessments can contribute to students' opportunities to learn important mathematics only if they are based on standards that reflect high expectations for all students. There can be no equity in assessment as long as excellence is not demanded of all. If we want excellence, the level of expectation must be set high enough so that, with effort and good instruction, every student will learn important mathematics.

We have much to learn about how to maintain uniformly high performance standards while allowing for assessment approaches that are tailored to diverse backgrounds. Uniform application of standards to a diverse set of tasks and responses poses an enormous challenge that we do not yet know how to do fairly and effectively. Nonetheless, the challenge is surely worth accepting.

OBSTACLES AND CHALLENGES

The boldness of our vision for mathematics assessment should not blind us to either the obstacles educators will face or the limitations on resources we possess for making it come about. Even if new assessments were to magically appear and be implemented across the nation, many substantial problems will remain. Examples of important, unresolved issues abound:

- Open-ended problems are not necessarily better than well-defined tasks. The mere labels "performance assessment" and "open ended" do not guarantee that a task meets sound educational principles. For example, open-ended problems can be interesting and engaging but mathematically trivial. Performance tasks can be realistic and mathematically appropriate but out of harmony with certain students' cultural backgrounds.

- The equity principle implies that students must be provided an opportunity to learn the mathematics that is

> Students cannot
> be assessed
> fairly on
> mathematics
> content that they
> have not had an
> opportunity to
> learn.

Our vision for

mathematics

assessment

should not blind

us to the

obstacles that

must be

overcome.

assessed and that schools must be held to "school delivery standards" to ensure that students are provided with appropriate preparation, particularly for any high-stakes assessment. However, many would argue that past remedies designed to improve schools often failed precisely because the emphasis was placed on the resources schools should provide rather than the outcomes that schools should achieve.

- The equity principle also requires some consideration of consequences for schools of the way assessments are used. Fair inferences can be drawn and comparisons can be made only when assessment data include information on the nature of the students served by the school, students' opportunities to learn the mathematics assessed, and the adequacy of resources available to the school. Assessments based only on partial data—typically outcome scores on basic skills—can seriously mislead the public about how schools are performing and how to improve them.

- On the job and in the real world, knowledge is frequently constructed and validated in group settings rather than through individual exploration. Mathematics is no exception: learning and performance are frequently improved in group settings. Hence assessment of learning must reflect the value of group interaction. The challenge of fairly appraising an individual's contribution to group efforts is immense, posing unresolved problems both for industry and education.

- New performance-based assessments introduce significant challenges both for the mathematical expertise of those who score assessments and for the guidelines used in scoring. Problem solving legitimately may involve some false starts or blind alleys; students whose work includes such things are doing important mathematics and their grades need to communicate this in an appropriate fashion. All graders must be alert to the unconventional, unexpected answer that, in fact, may contain insights that the assessor had not anticipated. Of course, the greater the chances of unanticipated responses, the greater the mathematical sophistication needed by those grading the tasks.

- As assessments become more complex and more connected to real-world tasks, there is a greater chance that the underlying assumptions and points of view may not apply equally to all students, particularly when differences in background and instructional histories are involved. Despite good intentions and best efforts to make new assessments fairer to all students than traditional forms of testing, preliminary research does not confirm the corollary expectation that group differences in achievement will diminish. Indeed, recent studies suggest that differences may be magnified when performance assessment tasks are used.

- Teachers are a fundamental key to assessment reform. As evaluation of student achievement moves away from short-answer recall of facts and algorithms, teachers will have to become skilled in using and interpreting new forms of assessment. As a result, teachers' professional development—at both the preservice and inservice levels—will become increasingly important.

- To the extent that communication is a part of mathematics, differences in communication skill must be seen as differences in mathematical power. To what extent are differences in ability to communicate to be considered legitimate differences in mathematical power?

- Current assessment frameworks, derived as they were from a measurement-based tradition largely divorced from mathematics itself, rarely conform to the principles of content, learning, and equity. Today's mathematics reveals the paramount importance of interconnections among mathematical topics and of connections between mathematics and other domains. Much assessment tradition, however, is based on an atomistic approach that hides connections both within mathematics and among mathematical and other domains.

BENEFITS FROM IMPROVED ASSESSMENT

Assessment based on the principles of content, learning, and equity are already being tested in numerous schools and jurisdictions in the United States. It is clear already that despite obstacles and challenges, many benefits accrue even beyond the central goal of improved assessment.

Assessments represent an unparalleled tool for communicating the goals and substance of mathematics education reform to various stakeholders. Assessments make the goals for mathematics learning real to students, teachers, parents, policymakers, and the general public, all of whom need to understand clearly where mathematics reform will take America's children and why they should support the effort. Assessments can be enormously helpful in this re-education campaign, especially if the context and rationale for various tasks are explained in terms that the public can understand.

Improved assessment can lead to improved instruction. Assessment can play a key role in exemplifying the new types of mathematics learning students must achieve. Assessments can indicate to students not only what they should learn but also the criteria that will be used in judging their performance. For example, a classroom discussion of an assessment in which students grade some (perhaps fictional) work provides a purely instructional use of an assessment device. The goal is not to teach answers to questions that are likely to arise, but to engage students in thinking about performance expectations.

Assessment can also be a powerful tool for professional development as teachers work together to understand new expectations and synchronize their expectations and grades. Teachers are rich sources of information about their students. With training on methods of scoring new assessments, teachers can become even better judges of student performance.

LOOKING TO TOMORROW

Improved assessment is not a panacea for the problems in mathematics education. Our findings neither diminish nor reject important, time-honored measurement criteria for evaluating assessment; nor do they suggest that changes in assessment alone will bring about education reform. Clearly, they will not.

Assessments represent an unparalleled tool for communicating the goals and substance of mathematics education reform to various stakeholders.

What we can say with assurance is that if old assessments remain in use, new curriculum and teaching methods will have little impact. Moreover, if new assessments are used as inappropriately as some old assessments, little good will come of changes in assessment.

It will take courage and vision to stay the course. As changes in curriculum and assessment begin to infiltrate the many jurisdictions of the U.S. educational system, these changes will at the outset increase the likelihood of mismatches among the key components of education: curriculum, teaching, and assessment. It is not unlikely that performance will decline initially if assessment reform is not tightly aligned with reform in curriculum and teaching.

Mathematics education is entering a period of transition in which there will be considerable exploration. Inevitably there will be both successes and failures. No one can determine in advance the full shape of the emerging assessments. Mathematics education is in this respect an experimental science, in which careful observers learn as much from failure—and from the unexpected—as from anticipated success. The necessary change will be neither swift nor straightforward. Nevertheless, we cannot afford to wait until all questions are resolved. It is time to put educational principles at the forefront of mathematics assessment.

> Although the necessary change in mathematics assessment will be neither swift nor straightforward, we cannot afford to wait until all questions are resolved.

1 A VISION OF SCHOOL MATHEMATICS

Mathematics is the key to opportunity. No longer just the language of science, mathematics now contributes in direct and fundamental ways to business, finance, health, and defense. For students, it opens doors to careers. For citizens, it enables informed decisions. For nations, it provides knowledge to compete in a technological community. To participate fully in the world of the future, America must tap the power of mathematics.

Communication has created a world economy in which working smarter is more important than merely working harder. Jobs that contribute to this world economy require workers . . . who are prepared to absorb new ideas, to adapt to change, to cope with ambiguity, to perceive patterns, and to solve unconventional problems. It is *these* needs, not just the need for calculation (which is now done mostly by machines), that make mathematics a prerequisite to so many jobs. More than ever before, Americans need to think for a living; more than ever before, they need to think mathematically.[1]

So opens the first chapter of *Everybody Counts: A Report to the Nation on the Future of Mathematics Education,* which describes a vision of the mathematics that should guide education so that students will work smarter and think more mathematically. The vision calls for changes in the mathematics taught, in the way it is taught, and in how it is assessed. Changes in mathematics assessment, the subject of this report, should be seen as one piece of the larger picture of reform in school mathematics.

Inside the classroom, teachers are working to change the mathematics they teach and how they teach it for many reasons, some of which they can find in their own classrooms. Far too many of their students—especially members of traditionally underserved groups—are turning away from mathematics, dropping out either physically or mentally.[2] Few students who stay with mathematics show much enthusiasm for it. It seems too abstract, too unrelated to either their present lives or their futures. Teachers are unhappy that students remember little of the mathematics they have been taught and seem incapable of using it.

Outside the classroom, politicians and school administrators, backed by the public, express dismay over low scores on mathematics achievement tests. They worry about deteriorating American competitiveness in international markets when students' mathematics skills seem to be declining.[3] They want teachers to teach more mathematics to more students while maintaining or increasing test scores. At the same time, teachers are being told by their professional associations that the mathematics they teach should be more applicable to life than is now common, that their teaching should generate active learning, and that their assessment of student learning be attuned not just to judging but to helping students learn.[4]

On the surface, the pressures to change mathematics instruction look inconsistent, with teachers caught in the middle. Nevertheless, all the pressures reflect disappointment with the lack of interest and accomplishment so many students show. The message is the same: School mathematics is out of step with today's world and is neither well taught nor well learned.

Three pivotal forces are moving mathematics teachers toward a different approach to their teaching. These forces are changing ideas about

> what should be taught,
> how it should be taught, and
> to whom it should be taught.

Motivating the first is a more comprehensive view of mathematics and its expanding role in society. Motivating the second is a resurgence of the view that mathematics must be made meaningful to students if it is to be learned, retained, and used. Motivating the third is the growing belief that all students can and should learn more mathematics.

CHANGES IN MATHEMATICS AND IN MATHEMATICS EDUCATION

Since 1900, the growth of the mathematical sciences—in scope and in application—has been explosive.[5] The last 40 years have been especially productive, as advances in high-speed computing have opened up new lines of research and new ways mathematics can be applied. Problems in economics, social science, and life science, as well as large-scale problems in natural science and engineering, used to be unapproachable through mathematics. Suddenly, with the aid of computers and the new tools provided by research, many of these problems have become accessible to mathematical analysis. Applications derived from data analysis and statistics, combinatorics and discrete mathematics, and information theory and computing have greatly extended the definition and reach of the mathematical sciences.

An explosion in the way mathematics is used in society mirrors the explosion in mathematics itself. Today we encounter uses of mathematics in every corner of our lives. Graphs, charts, and statistical data appear on television and in newspapers. The results of opinion polls are reported along with their margins of error. Lending institutions advertise variously computed interest rates for loans. We listen to music composed and performed with the aid of computers, and we watch the fantastically detailed pictures of imaginary worlds that computers draw. Computers also do a host of ordinary tasks. They scan bar codes on purchases, keep track of inventories, make travel reservations, and fill out income tax forms. The citizen's need to perform simple calculations may have decreased, but there has been a dramatic increase in the need to interpret, evaluate, and understand quantitative information presented in a variety of contexts.

Although some people do not need or use highly technical mathematics in their daily jobs, many others do. The complexity of daily life requires that we all be able to reason with numbers. Any car or home buyer ought to understand how interest rates work even though a computer may be doing the calculation. Anyone building a house or redecorating a room should be able to make and read a scale drawing. Newspaper readers and television viewers

The mathematics taught in school must change in support of the way mathematics is used in our society.

should be able to draw correct inferences from data on social problems such as pollution, crime, drugs, and disease. Buyers of insurance and purchasers of stock need to know something about the calculation of risk. All adults should be both able and disposed to use mathematics to make practical decisions, to understand public policy issues, to do their job better, to enhance their leisure time, and to understand their culture.[6] Mathematics and the ways it is used are changing. The mathematics taught in school must change in support of the way mathematics is used in our society.

WHAT MATHEMATICS SHOULD BE LEARNED

Over the years, professionals concerned with mathematics education have developed a coherent view of what mathematics is important, despite some disagreements along the way. At the turn of this century, the mathematician E. H. Moore called for a refurbished school mathematics curriculum. He expressed the hope that twentieth century students would at least encounter "in thoroughly concrete and captivating form, the wonderful new notions of the seventeenth century,"[7] particularly an introduction to calculus. He wanted primary school children to make models and study intuitive geometry along with arithmetic and algebra. He argued for tight connections and a blurring of the distinctions between all parts of school mathematics but especially between its pure and applied sides.

The prestigious National Committee on Mathematical Requirements of the Mathematical Association of America, reporting in 1923, formulated the aims of mathematical instruction as practical, disciplinary, and cultural. They viewed the idea of relationship or dependence, which can be expressed in the mathematical concept of function, as encompassing many of the disciplinary aims. They also deemed fundamental an appreciation of the power of mathematics. In words that have a contemporary ring, they said

> The primary purpose of the teaching of mathematics should be to develop those powers of understanding and of analyzing relations of quantity and of space which are necessary to an insight into and control over our environment and to an appreciation of the progress of civilization in its various aspects, and to develop those habits of thought and of action which will make these powers effective in the life of the individual.[8]

Launching the movement that became known as "the new math," the College Board's Commission on Mathematics in 1959 made the case for revamping the school mathematics curriculum:

> The traditional curriculum fails to reflect adequately the spirit of contemporary mathematics, which seeks to study all possible patterns recognizable by the mind, and by so striving has tremendously increased the power of mathematics as a tool of modern life. Nor does the traditional curriculum give proper emphasis to the fact that the developments and applications of mathematics have always been not only important but indispensable to human progress.[9]

Recently, in its *Curriculum and Evaluation Standards for School Mathematics*, the National Council of Teachers of Mathematics (NCTM) contended that the traditional sequence of mathematics courses leading to the calculus is inadequate:

> Students should be exposed to numerous and varied interrelated experiences that encourage them to value the mathematical enterprise, to develop mathematical habits of mind, and to understand and appreciate the role of mathematics in human affairs; that they should be encouraged to explore, to guess, and even to make and correct errors so that they gain confidence in their ability to solve complex problems; that they should read, write, and discuss mathematics; and that they should conjecture, test, and build arguments about a conjecture's validity.[10]

The *Standards* delineate the mathematics students need to learn under various headings, some familiar (measurement, algebra, probability, problem solving), some perhaps less so (communication, spatial sense, discrete mathematics). In an echo of the 1923 report, the *Standards* emphasize mathematical power and outline experiences designed to help all students gain that power. Particularly important are the processes of mathematical thinking whereby students learn problem solving, communicating, reasoning, and making connections. Concurrent with the dramatic changes in our society over the last century—including the revolution in information technology and the recent increase in economic competitiveness—the profession's view of what mathematics is important is evolving consistently. As these statements from mathematics educators show, the profession has long sought to move instruction beyond a narrow focus on calculation to a deeper consideration of the meaning, process, and uses of mathematics. Criticism has been aimed at the so-called traditional curriculum, with its stress on symbol manipulation, its fragmentation, and its artificial treatment of applications.

Today's consensus on mathematics learning embraces several important components. Students should be involved in finding, making, and describing patterns. They should construct mathematical models for both practical and theoretical situations— using technology when appropriate—learning to represent and reason about quantities and shapes, to devise and solve challenging problems, and to communicate what they have learned. Students also should encounter mathematics as a human endeavor, learning something of its history in various cultures, coming to appreciate its aesthetic side, and understanding its role in contemporary society and its connections to other disciplines and areas of knowledge.

School mathematics from kindergarten to 12th grade should offer much more than procedural skills. It should equip students not only for the further study of mathematics and other subjects but also to use mathematics creatively and effectively in their daily lives and subsequent careers. Elementary school mathematics should go well beyond computational arithmetic, which is only one aspect of mathematics. Instead, it should also include topics such as number sense, estimation, and an introduction to geometry, probability, statistics, and algebra, all treated in ways that deemphasize the boundaries between these strands and developed through activities that use physical objects. Middle and high school mathematics should continue the development of the strands begun in elementary school mathematics and in addition should include combinatorics, discrete mathematics, logic, number theory, trigonometry, and some basic ideas from calculus. Fundamental mathematical structures (relations, functions, operational systems) should guide the selection of topics and serve as unifying themes.

Students learn important mathematics when they are using mathematics in relevant contexts that require them to apply what they know and to extend how they think. The context may be fanciful or it may resemble the real world, but the content should make sense to students and involve mathematics they need to know.

Students learn important mathematics when they are using it in relevant contexts that require them to apply what they know and to extend how they think.

How Mathematics Should Be Taught

The phenomenal growth in mathematics and the way it now permeates our world require a fresh look at what it means to understand and do mathematics. The advent of powerful computing technology has made mathematics, more than ever before, an experimental science, with the same need for observation and inquiry. *Reshaping School Mathematics: A Philosophy and Framework for Curriculum*, from the Mathematical Sciences Education Board (MSEB), puts it as follows:

> Mathematics is a science. Observations, experiment, discovery, and conjecture are as much part of the practice of mathematics as of any natural science. Trial and error, hypothesis and investigation, and measurement and classification are part of the mathematician's craft and should be taught in school.[11]

Moreover, the availability of computers has renewed the emphasis on realistic applications, greatly simplifying the treatment of data in the classroom and permitting dynamic representations of complex processes. With the aid of computers, students can have experiences heretofore impossible in representing patterns, estimating solutions, and exploring how changes in one representation affect another. Technology such as powerful hand-held graphing calculators—in reality, hand-held computers—allow real-life problems to be explored in the classroom in all their complexity.

Just as when they study the natural, physical, and social sciences, students of mathematics should be given opportunities to pose problems and advance hypotheses after they have examined a situation for the patterns and relationships it contains. They need to learn how to construct and use mathematical models of real phenomena. They should be taught to make and test their inferences, using estimates and the mathematics of uncertainty as well as the more familiar techniques of arithmetic, algebra, geometry, and calculus. Such activities will help them understand both how their model works and how that model falls short of capturing the complexity of the situation.

Students learn

best and most

enduringly by

reflecting on

their experience

and by

communicating

with others

about it.

Mathematics is a science, "the science of patterns." [12] It enables scientists to analyze the regularities in their data and fourth graders to understand and make sense of the multiplication table. More than that, mathematics is also a language, the language of much of today's business and commerce as well as "the language in which nature speaks." [13]

To teach mathematics as both a useful science and a living language, rather than simply as a collection of arbitrary rules to be memorized, demands a different approach to the subject. The interrelation of mathematical ideas from all branches of the mathematical sciences needs to be stressed from kindergarten through high school and beyond. For more than a century, educators have bemoaned the compartmentalization of school mathematics into arithmetic, algebra, trigonometry, and so on.[14] If their mathematics is to be both understandable and usable, students must learn to apply ideas from algebra and geometry together with statistics and discrete mathematics to the analysis of data. They can then use this analysis to pose problems, test hypotheses, construct mathematical models, and communicate their findings. In such activities, it is pointless to maintain a separation between mathematical topics and contradictory to the process of mathematical exploration.

Good teachers have long recognized that mathematics comes alive for students when it is learned through experiences they find meaningful and valuable. Students want to make sense of their world. Mathematics becomes part of that world when it is seen both as sensible in itself and as a tool for making sense out of otherwise confusing situations. Research from cognitive science and instructional psychology supports the view that successful learners build their own understanding of subject matter. Much of this research uses mathematics as a discipline for exploring issues of learning.[15] According to this research, even the youngest learners take nothing ready-made; instead, they filter what they learn through their own sensibilities and through what they already know. Students learn best and most enduringly by reflecting on their experience and by communicating with others about it.

A new view of mathematical performance is developing in which the focus is on the concrete tasks students perform in a specific social context rather than on abstract abilities that students are assumed to possess.[16] Learners benefit from performing a challenging task and developing their understanding of it in interaction with others. Learning mathematics along with others helps students develop confidence in doing mathematics and a positive disposition toward the subject.[17]

Who Should Learn Mathematics

Our curriculum is still organized in ways that prevent many students from gaining access to the mathematics they need. In America, we have developed a two-tiered system in which "poor and minority students are underrepresented in college-preparatory classes such as algebra and geometry and overrepresented in dead-end classes such as consumer math and general math."[18] Research has demonstrated that practices of ability grouping for instruction deny many students the opportunity to learn valuable mathematics.[19] Further complicating the problem is that many classes—including those labeled college prep—do not provide the mathematics education needed by today's students. Just gaining access to such classes is not enough. All students must learn important mathematics in these classes.

Americans have tended to view achievement in mathematics as a product of special talent rather than effort.[20] Yet most young children like mathematics and see it as something they are capable of doing. Indeed, most children enter school with quite sophisticated theories about how numbers are used in their world. In elementary school, however, mathematics begins to look different to them. Students judge mathematics as harder to learn than other kinds of content, see themselves as less capable of learning mathematics on their own, and feel more dependent on direct instruction from teachers and others.[21]

These views stem in large part from experiences the students have in school. Too much of students' mathematics instruction is divorced from any context that connects with their lives and with the intuitive understanding they bring with them to class.[22] These views are also shaped and supported by the attitudes of the adults around them: Many parents who would find it completely unacceptable if their children did not learn to read are content to accept and excuse low performance in mathematics.

Evidence is rapidly accumulating that the challenging levels of mathematics needed for the future can be learned by all students.

Fortunately, evidence is rapidly accumulating that the challenging levels of mathematics needed for the future can be learned by all students.[23] From elementary school to college, programs are emerging that reduce drop-out rates and, more importantly, record significant gains by traditionally underrepresented students in challenging mathematics classes.[24] A report from the Algebra Project stated that

> The best example is the King School [in Cambridge, MA], where the program has been in place for 10 years. Before the Algebra Project, few students took the optional advanced-placement qualifying test in ninth grade, and virtually none passed. By 1991, the school's graduates ranked second in Cambridge on the test.[25]

Such programs invalidate the myth that only a talented few can learn important mathematics. Results from studies around the world, particularly from countries with high achievement in mathematics, also invalidate the myth. In many Asian countries, for example, academic standards are high, everyone has high expectations for all children, and people believe that all children can learn to those standards if they are taught well and work hard.[26]

To produce an adult population with the knowledge it will need, mathematics education must reflect and support the view that all students can learn significantly more mathematics than is currently the case. Assessment plays a critical role in this process because assessment will measure and influence students' learning.

CURRENT EFFORTS AT REFORM

Efforts to reform school mathematics are proceeding along three lines: revitalizing the curriculum, redesigning the professional development of teachers, and reconceptualizing the assessment of learning. Strategies aimed at curriculum and professional development are set out in some detail in recent reports that include *Everybody Counts*, *Curriculum and Evaluation Standards for School Mathematics*, *Reshaping School Mathematics*, *Professional Standards for Teaching Mathematics*, *A Call for Change*, and *Counting on You*.[27] These reports agree that goals for student performance are shifting from a narrow focus on routine skills to the provision of a variety of experiences aimed at developing students' mathematical power. They encourage the movement from teaching as the transmission of knowledge to teaching as the stimulation of learning. They view teachers as the central agents for changing school mathematics and ask that teachers be given continuing support and adequate resources.

Learning as sense-making and teaching as providing experiences in which sense can be made are at the crux of the vision of school mathematics emerging today in American society.[28] Yet, these ideas are far from new. Thinkers as diverse as Aristotle, Dewey, and Piaget or as similar as Pestalozzi, Froebel, and Pólya have all expressed such thoughts. Mathematics teachers have for centuries found it difficult to lead students to a deep understanding of how and why mathematics works as it does. What is different now that makes successful reform in mathematics education more likely?

Part of the answer can be found in the reports noted above, as they document how efforts are moving ahead together, for perhaps the first time, on three fronts—curriculum, professional development, and assessment—to ensure the necessary transformation of mathematics learning. Another part of the answer can be found in the widespread consensus that change in assessment is critical to improving education. Content and measurement experts alike have been exploring ways of creating assessments that promote and support educational reform. Until recently, however, there was little collaboration and very few points of cross-fertilization between the two fields on how this might be accomplished. This picture is changing. As mathematics experts are grappling with educational principles to guide assessment, so measurement experts are re-examining the criteria by which the technical quality of assessments is evaluated. Like new views of mathematics teaching and learning, new technical criteria and procedures for making them operational are being refined at the present time.[29]

Reform in mathematics assessment must be based not simply on what is easy to assess but much more importantly on what needs to be assessed.

ENDNOTES

1 National Research Council, Mathematical Sciences Education Board, Board on Mathematical Sciences, and Committee on the Mathematical Sciences in the Year 2000, *Everybody Counts: A Report to the Nation on the Future of Mathematics Education* (Washington, D.C.: National Academy Press, 1989), 1.

2 Ibid., 17-29.

3 National Commission on Excellence in Education, *A Nation at Risk: The Imperative for Educational Reform* (Washington, D.C.: U.S. Government Printing Office, 1983); U.S. Department of Education, *America 2000: An Education Strategy Sourcebook* (Washington, D.C.: U.S. Government Printing Office, 1991); U.S. Department of Labor, Secretary's Commission on Achieving Necessary Skills, *What Work Requires of Schools: A SCANS Report for America 2000* (Washington, D.C.: U.S. Department of Labor, 1991); James G. Glimm, ed., *Mathematical Sciences, Technology, and Economic Competitiveness* (Washington, D.C.: National Academy Press, 1991).

4 National Council of Teachers of Mathematics, *Curriculum and Evaluation Standards for School Mathematics* (Reston, VA: Author, 1989); National Council of Teachers of Mathematics, *Professional Standards for Teaching Mathematics* (Reston, VA: Author, 1991).

5 *Everybody Counts,* 34; *See also* National Research Council, Committee on Resources for the Mathematical Sciences, *Renewing U.S. Mathematics: Critical Resource for the Future* (Washington, D.C.: National Academy Press, 1984).

6 *Everybody Counts,* 32-33.

7 Eliakim H. Moore, "On the Foundations of Mathematics," presidential address to the ninth annual meeting of the American Mathematical Society, *Science,* 13 March 1903.

8 Mathematical Association of America, National Committee on Mathematical Requirements, *The Reorganization of Mathematics in Secondary Education* cited in National Council of Teachers of Mathematics, *The Teaching of Secondary School Mathematics,* NCTM Yearbook (Washington, D.C.: Author, 1970).

9 College Entrance Examination Board, Commission on Mathematics, *Program for College Preparatory Mathematics* (New York, NY: Author, 1959).

10 *Curriculum and Evaluation Standards for School Mathematics,* 5.

11 National Research Council, Mathematical Sciences Education Board, *Reshaping School Mathematics: A Philosophy and Framework for Curriculum* (Washington, D.C.: National Academy Press, 1990), 10-11.

12 Lynn Arthur Steen, "The Science of Patterns," *Science,* 29 April 1988, 611-616.

13 *Reshaping School Mathematics,* 11.

14 Thomas Hill, *The True Order of Studies* (New York, NY: G. P. Putman's Sons, 1876); *See also* Solberg E. Sigurdson, "The Development of the Idea of Unified Mathematics in the Secondary School Curriculum 1890-1930" (Ph.D. diss., University of Wisconsin, 1962).

15 Two classroom-based research projects in mathematics have provided

some strong evidence supporting the constructivist approach: The first is the "Second Grade Classroom Teaching Project" that has been reported extensively by Paul Cobb, Terry Wood, Erna Yackel and their colleagues [see, for example, Cobb, Wood, and Yackel "Classrooms as Learning Environments for Teachers and Researchers," in Robert Davis, Carolyn Maher and Nel Noddings, eds., *Constructivist Views on the Teaching and Learning of Mathematics*, monograph, no. 4 (Reston, VA: National Curriculum of Teachers of Mathematics, 1990), 125-146]. The second is the Cognitively Guided Instruction project, which has been described by Elizabeth Fennema, Thomas Carpenter, and Penelope Peterson [see, for example, Fennema, Carpenter, and Peterson, "Learning Mathematics with Understanding: Cognitively Guided Instruction," in J. Brophy, ed., *Advances in Research in Teaching* (Greenwich, CT: JAI Press, 1989), 195-221]. For a more general discussion, see Lauren B. Resnick, *Education and Learning to Think*, National Research Council, Committee on Mathematics, Science, and Technology Education, Commission on Behavioral and Social Sciences and Education (Washington, D.C.: National Academy Press, 1987); and *Everybody Counts,* 58-59.

[16] Yvette Solomon, *The Practice of Mathematics* (London, England: Routledge, 1989), 179-187.

[17] Thomas L. Good, Catherine Mulryan, and Mary McCaslin, "Grouping for Instruction in Mathematics: A Call for Programmatic Research on Small-Group Processes," in Douglas A. Grouws, ed., *Handbook of Research on Mathematics Teaching and Learning* (New York, NY: Macmillan, 1992), 165-196. Neil Davidson and Toni Worshem, "Enhancing Thinking Through Cooperative Learning" (New York, NY: Teachers College, 1992).

[18] Vinetta Jones, "Responses to Three Questions" (Paper prepared for The State of American Public Education: Views on the State of Public Schools, Washington, D.C., 4-5 February 1993).

[19] Jeannie Oakes, *Keeping Track: How Schools Structure Inequality* (New Haven, CT: Yale University Press, 1985); Jeannie Oakes, *Multiplying Inequalities: The Effect of Race, Social Class, and Tracking on Opportunities to Learn Mathematics* (Santa Monica, CA: Rand Corporation, 1990); Leigh Burstein, ed., *Student Growth and Classroom Processes*, vol. 3, *IEA Study of Mathematics* (Oxford, England: Pergamon Press, 1992).

[20] Harold W. Stevenson and James W. Stigler, *The Learning Gap: Why Our Schools Are Failing and What We Can Learn from Japanese and Chinese Education* (New York, NY: Summit Books, 1992), 113-129.

[21] S. Stodolsky, S. Salk, and B. Glaessner, "Student Views About Learning Math and Social Studies," *American Educational Research Journal* 28:1 (1991), 89-116.

[22] Herbert Ginsburg, *Children's Arithmetic: The Learning Process* (New York, NY: D. Van Nostrand Co., 1977); Lauren B. Resnick and Wendy W. Ford, *The Psychology of Mathematics for Instruction* (Hillsdale, NJ: Lawrence Erlbaum Associates, 1981).

[23] Michael Cole and Peg Griffin, eds., *Contextual Factors in Education: Improving*

Science and Mathematics Education for Minorities and Women (Madison, WI: Wisconsin Center for Education Research, 1987); Edward A. Silver, Margaret S. Smith, and Barbara S. Nelson, "The QUASAR Project: Equity Concerns Meet Mathematics Education Reform in the Middle School," to appear in Walter G. Secada, Elizabeth Fennema, and Lisa Byrd, eds., *New Directions in Equity for Mathematics Education,* (Draft version, April 1993); Deborah A. Carey et al., "Cognitively Guided Instruction: Towards Equitable Classrooms" to appear in *New Directions in Equity for Mathematics Education.*

[24] Cynthia M. Silva and Robert P. Moses, "The Algebra Project: Making Middle School Mathematics Count," *Journal of Negro Education* 59:3 (1990), 388; P. Uri Treisman, "Studying Students Studying Calculus: A Look at the Lives of Minority Mathematics Students in College," *College Mathematics Journal* 23:5 (1992), 362-372.

[25] Alexis Jetter, "Mississippi Learning," *The New York Times Magazine*, 21 Feb 1993, 35.

[26] This case is made most extensively by Harold W. Stevenson and James W. Stigler in *The Learning Gap: Why Our Schools are Failing and What We Can Learn from Japanese and Chinese Education.*

[27] The Mathematical Association of America, *A Call for Change: Recommendations for the Mathematical Preparation of Teachers of Mathematics*, (Washington, D.C.: Author, 1991); The National Research Council, Mathematical Sciences Education Board, *Counting on You* (Washington, D.C.: National Academy Press, 1991). See endnotes 1, 4, and 11 for additional citations.

[28] *Curriculum and Evaluation Standards for School Mathematics*, 15-19, 38-40; *Everybody Counts*, 46-48, 61.

[29] See Lee J. Cronbach, "Five Perspectives on Validity Argument," in Howard Warner and Henry I. Baum, *Test Validity* (Hillsdale, NJ: Lawrence Erlbaum Associates, 1988), 3-17; Robert L. Linn, Eva L. Baker, and Stephen B. Dunbar, "Complex, Performance-Based Assessment: Expectations and Validation Criteria," *Educational Researcher* 20:8 (1991), 15-21; John R. Frederiksen and Allan Collins, "A Systems Approach to Educational Testing," *Educational Researcher* 18:9 (1989), 27-32; Samuel Messick, "Validity," in R.L. Linn, ed., *Educational Measurement* (New York, NY: American Council on Education/ Macmillan, 1989), 13; Pamela Moss, "Shifting Conceptions of Validity in Educational Measurement" (Paper presented at the annual meeting of AERA, San Francisco, April 1992.); Lorrie Shepard, "Psychometricians' Beliefs about Learning," *Educational Researcher* 20:6 (1991), 33-42; Eva L. Baker and Robert L. Linn, "The Technical Merit of Performance Assessments," *The CRESST Line*, Newsletter of the UCLA Center for Research on Evaluation, Standards, and Student Testing, Spring 1993, 1); Stephen B. Dunbar, Daniel Koretz, and H.D. Hoover, "Quality Control in Development and Use of Performance Assessments," *Applied Measurement in Education* 4:4 (1991), 289-304; Eva L. Baker, Harold F. O'Neil, and Robert L. Linn, *What Works in Alternative Assessment?* (Draft version, September 1992); Stephen B. Dunbar and Elizabeth A. Witt, "Design Innovations in Measuring Mathematics Achievement" (Paper commissioned by the Mathematical Sciences Education Board, September 1993, appended to this report).

2 A VISION OF MATHEMATICS ASSESSMENT

ssessment is the means by which we determine what students know and can do. It tells teachers, students, parents, and policymakers something about what students have learned: the mathematical terms they recognize and can use, the procedures they can carry out, the kind of mathematical thinking they do, the concepts they understand, and the problems they can formulate and solve. It provides information that can be used to award grades, to evaluate a curriculum, or to decide whether to review fractions. Assessment can help convince the public and educators that change is needed in the short run and that the efforts to change mathematics education are worthwhile in the long run. Conversely, it can thwart attempts at change. Assessment that is out of synchronization with curriculum and instruction gives the wrong signals to all those concerned with education.

Mathematics assessments are roughly divided into two categories: internal and external. Internal assessments provide information about student performance to teachers for making instructional decisions. These assessments may be for high or low stakes, but they exert their chief influence within the walls of the classroom. External assessments provide information about mathematics programs to state and local agencies, funding bodies, policymakers, and the public. That information can be used either to hold program managers accountable or to monitor the program's level of performance. These assessments are used primarily by people outside the immediate school community. Although internal assessment is perhaps more obviously and directly connected with the improvement of mathematics learning than external assessment, both types of assessment should advance mathematics education.

THE ROLE OF ASSESSMENT IN REFORM

Assessment can play a powerful role in conveying, clearly and directly, the outcomes toward which reform in mathematics is aimed. As assessment changes along with instruction, it can help teachers and students keep track of their progress toward higher standards. Many reformers see assessment as much more than a signpost, viewing it as a lever for propelling reform forward.[1] It is essential that mathematics assessment change in ways that will both support and be consistent with other changes under way in mathematics education.

From their beginnings in the last century, standardized achievement tests have been used in American schools not only to determine what students have learned but also to induce better teaching. The written tests administered by the Boston School Committee in 1845 led to rankings of schools by level of achievement and to recommended changes in instructional methods.[2] The New York State Regents Examinations were set up primarily to maintain standards by showing teachers what their students needed to know.[3] The traditional view of many Americans that tests and examinations can do more than measure achievement is reflected in this quotation from a 1936 book on assessment prepared for the American Council on Education: "Recently increasing emphasis has been placed upon examinations as means for improving instruction, and as instruments for securing information that is indispensable for the constructive educational guidance of pupils." [4]

Researchers are beginning to document more thoroughly the effects of assessment, determining, in effect, whether this traditional view is justified. A 1992 study by the Center for the Study of Testing, Evaluation, and Educational Policy at Boston College examined the content of the most commonly used tests embedded in textbooks and standardized tests in mathematics and science for grades 4 to 12 and how they influence instruction. The authors noted that the tests fell far short of the reform vision and concluded that

> Since textbook tests were found to be similar to standardized tests in the skills they measure, and since these tests are widely used, an emphasis on low level thinking skills extends beyond the instructional time spent preparing for state and district mandated standardized tests. The tests most commonly taken by stu-

Mathematics

assessment

must change in

ways that will

both support

and be

consistent with

other changes

under way in

mathematics

education.

dents—both standardized tests and textbooks tests—emphasize and mutually reinforce low level thinking and knowledge, and were found to have an extensive and pervasive influence on math and science instruction nationwide.[5]

Proponents of mathematics education reform have expressed the view that the goal of more and better mathematics learning for all students cannot be realized if assessment remains wedded to what is easy to measure and what has traditionally been taught. The messages sent by new views of content, teaching, and learning will be contradicted by the values that such assessment practices communicate. Some teachers may attempt to respond to both messages by preparing students for the tests while, simultaneously, trying to offer students some taste of a richer, more challenging curriculum. Other teachers may continue to teach as they have always taught, reasoning that the tests used to make important decisions about students' lives, teachers' salaries, and educational quality indicate what is *truly* important for students to learn. If current assessment practices prevail, reform in school mathematics is not likely to succeed.

If current assessment practices prevail, reform in school mathematics is not likely to succeed.

Suppose assessment practice were to change in American mathematics classes. What if, at the end of a unit, students wrote an essay explaining how two situations could both be modeled with the same exponential function, instead of being tested on skills such as solving equations and choosing among definitions? Imagine students being assessed not only with a final examination taken in class but also on how well they could conduct and report a group investigation of the symmetries in a wallpaper pattern. Suppose students were allowed to use calculators, computers, and other resources on most tests and examinations, including those administered by external agencies. If such changes were to occur, many mathematics teachers would shift their instruction to prepare their students adequately for such assessments.

Reformers have proposed a host of innovative approaches to assessment, many of which are described in subsequent sections of this report. Leaders in the educational policy community are joining the chorus, arguing that minimum competence tests and basic skill assessments, like those commonly seen today, work against efforts to improve schools.[6] Low-level tests give coarse and deceptive readings of educational progress. Worse, they send the wrong message about what is important.[7] Assessments need to record genuine accomplishments in reasoning and in formulating and solving mathematical problems, not feats of rote memorization or proficiency in recognizing correct answers.[8]

The content of assessment needs to change along with its form if the vision of mathematics teaching and learning is to be attained. Portfolios of mathematical work can contribute to better teaching and learning only if the collections reflect work on meaningful tasks that require use of higher-level mathematical processes. Write-ups of mathematical investigations can support the vision only if the mathematics addressed is important rather than trivial. New forms of assessment can support efforts to change instruction and curriculum only if constructed in ways that reflect the philosophy and the substance of the reform vision of school mathematics. Obviously, changing the content and forms of assessment will not be sufficient to bring about reform. Such changes will have meaning only if curriculum changes and professional development for teachers are attended to as part of the reform process.

PRINCIPLES FOR ASSESSING MATHEMATICS LEARNING

In this chapter, three educational principles based on content, learning, and equity are set forth to guide changes in mathematics assessment. Underlying these three principles is the fundamental premise that assessment makes sense only if it is in harmony with the broad goals of mathematics education reform.

THE CONTENT PRINCIPLE
Assessment should reflect the mathematics that is most important for students to learn.

Any assessment of mathematics learning should first and foremost be anchored in important mathematical content. It should reflect topics and applications that are critical to a full understanding of mathematics as it is used in today's world and in students' later lives, whether in the workplace or in later studies. Assessments should reflect processes that are required for doing mathematics: reasoning, problem solving, communication, and connecting ideas. Consensus has been achieved within the discipline of mathematics and among organizations representing mathematics educators and teachers on what constitutes important mathematics. Although such consensus is a necessary starting point, it is important to obtain public acceptance of these ideas and to preserve local flexibility to determine how agreed-upon standards are reflected in assessments as well as in curricula.

As uses of mathematics change over time, visions of school mathematics and assessment must evolve in consonant ways. No existing conception of important content should constitute an anchor, preventing changes in assessment that are warranted by changing times. Thus, assessment development will require more significant collaboration between content and measurement experts than has been characteristic in the past. The goal of the content principle is to ensure that assessments are based on well-reasoned conceptions of what mathematics students will need to lead fully informed lives. Only if the mathematics assessed is important can the mathematics be justified as significant and valuable for students to know, and the assessment justified as supportive of good instruction and a good use of educational resources.

THE LEARNING PRINCIPLE
Assessment should enhance mathematics learning and support good instructional practice.

Although assessments can be undertaken for various purposes and used in many ways, proponents of standards-based assessment reform have argued for the use of assessments that contribute very directly to student learning. The rationale is that challenging students to be creative and to formulate and solve problems will not ring true if all students see are quizzes, tests, and examinations that dwell on routine knowledge and skill. Consciously or unconsciously, students use assessments they are given to determine what others consider to be significant.

There are many ways to accomplish the desired links between assessment and learning. Assessment tasks can be designed so that they are virtually indistinguishable from good learning tasks by attending to factors that are critical to good instructional design: motivation, opportunities to construct or extend knowledge, and opportunities to receive feedback and revise work. Assessment and instruction can be combined, either through seamlessly weaving the two kind of activities together or by taking advantage of opportuni-

> Assessment
> makes sense
> only if it is in
> harmony with
> the broad goals
> of mathematics
> education
> reform.

ties for assessment as instruction proceeds. Assessments can also be designed in ways that help communicate the goals of learning and the products of successful learning. In each of these approaches, the teacher's role is critical both for facilitating and mediating learning.

THE EQUITY PRINCIPLE
Assessment should support every student's opportunity to learn important mathematics.

The equity principle aims to ensure that assessments are designed to give every student a fair chance to demonstrate his or her best work and are used to provide every student with access to challenging mathematics.

Equity requires careful attention to the many ways in which understanding of mathematics can be demonstrated and the many factors that may color judgments of mathematical competence from a particular collection of assessment tasks.

Equity also requires attention to how assessment results are used. Often assessments have been used inappropriately to filter students out of educational opportunity. They might be used instead to empower students: to provide students the flexibility needed to do their best work, to provide concrete examples of good work so that students will know what to aim for in learning, and to elevate the students' and others' expectations of what can be achieved.

Equity also requires that policies regarding use of assessment results make clear the schools' obligations to educate students to the level of new content and performance standards.

. . . .

EDUCATIONAL PRINCIPLES IN CONTEXT

Time spent on assessment is increasing in classrooms across the country.[9] Separate assessments are often administered to answer a wide array of questions, from what the teacher should emphasize in class tomorrow to what the school system should do to improve its overall mathematics program. Whether the sheer number of assessments is reduced is not the primary issue. What is more critical is that any time spent on assessment be time used in pursuit of the goal of excellent education.

The principles described above provide criteria that aim to ensure that assessments foster the goal of excellent mathematics education. For decades, educational assessment in the United States has been driven largely by practical and technical concerns rather than by educational priorities. Testing as we know it today arose because very efficient methods were found for assessing large numbers of people at low cost. A premium was placed on assessments that were easily administered and that made frugal use of resources. The constraints of efficiency meant that mathematics assessment tasks could not tap a student's ability to estimate the answer to an arithmetic calculation, construct a geometric figure, use a calculator or ruler, or produce a complex deductive argument.

A narrow focus on technical criteria—primarily reliability—also worked against good assessment. For too long, reliability meant that examinations composed of a small number of complex problems were devalued in favor of tests made up of many short items. Students were asked to perform large numbers of smaller tasks, each eliciting information on one facet of their understanding, rather than to engage in complex problem solving or modeling, the mathematics that is most important.

In the absence of expressly articulated educational principles to guide assessment, practical and technical criteria have become de facto ruling principles. The content, learning, and equity principles are proposed not to challenge the importance of these criteria, but to challenge their dominance and to strike a better balance between educational and measurement concerns. An increased emphasis on validity—with its attention to fidelity between assessments, high-quality curriculum and instruction, and consequences—is the tool by which the necessary balance can be achieved.

In attempting to strike a better balance between educational and measurement concerns, many of the old measurement questions must be re-examined. For example, standardization has usually been taken to mean that assessment procedures and conditions are the same for every student. But from the perspective of fairness and equity, it might be more critical to assure that every student has the same level of understanding about the context and requirements of an assessment or task. The latter interpretation requires that some accommodation be made to differences among learners. For example, the teacher or examination proctor might be allowed to

> The content, learning, and equity principles challenge the dominance, not the importance, of traditional measurement criteria.

explain instructions when needed, a procedure that would be proscribed under prevailing practices. Standardization will remain important, but how it is viewed and how it is operationalized may require rethinking, as the new principles for assessment are put in place.[10]

To strike a

better balance

between

educational and

measurement

concerns,

many of the old

measurement

questions must

be re-examined.

Putting the content, learning, and equity principles first will present different kinds of challenges for different constituencies. It will mean finding new approaches for creating, scoring, and evaluating mathematics assessments. Some new approaches are being tried in schools today. Techniques are being developed that allow students to show what they know and can do and not simply whether they recognize a correct answer when they see one. These changes imply new roles for teachers. Much of the impulse behind the movement toward standardized testing over this century arose from a mistrust of teachers' ability to make fair, adequate judgments of their students' performance.[11] Teachers will have to be accorded more professional credibility as they are given increased responsibility for conducting and evaluating student responses on assessments developed to meet the three principles. Teachers will need systematic support in their efforts to meet these new professional responsibilities and challenges.

The principles also present challenges for assessment developers and researchers. Some issues that need clarification relate to the broader definitions of important content now embraced by the mathematics education community. Processes such as communication and reasoning, for example, previously have been classified as nonmathematical skills. Broadening the domain of important mathematics to include these skills may make it difficult to separate general cognitive skills from the outcomes of mathematics instruction, which may undermine validity as it is traditionally understood.[12]

Other open technical issues relate to the difficulty of establishing that assessment tasks actually evoke the higher-order processes they were designed to tap.[13] The array of solutions to high-quality mathematics tasks is potentially so rich that expert judgements will not be sufficient. Students may need to be interviewed about their solution approaches during or at the conclusion of a task. Student work also will need to be examined. A number of researchers are exploring different approaches for making process determinations from student work. As yet, however, there are no well established procedures for translating this kind of information into forms that are useful for evaluating how well assessments meet the content and learning principles or how well they satisfy the more traditional criterion of content validity.

Reordering priorities so that these new principles provide a foundation on which to develop new assessments puts student learning of mathematics ahead of other purposes for assessment. It is bound to have dramatic implications for mathematics assessment, not all of which can be foreseen now. The purpose of the remainder of this report, however, is to examine what is known from research, what questions still await answers, and what the wisdom of expert practice suggests about the principles and their implementation.

ENDNOTES

[1] Peter W. Airasian, "Symbolic Validation: The Case of State-Mandated, High-Stakes Testing," *Educational Evaluation and Policy Analysis* 10:4 (1988), 301-313; Eva L. Baker and Regie Stites, "Trends in Testing in the USA," in Susan H. Fuhrman and Betty Malen, eds., *The Politics of Curriculum and Testing*, 1990 Yearbook of the Politics of Education Association (London, England: The Falmer Press, 1990), 139-157; National Commission on Testing and Public Policy, *From Gatekeeper to Gateway: Transforming Testing in America* (Chestnut Hill, MA: Boston College, 1990); Grant Wiggins, "Teaching to the (Authentic) Test," *Educational Leadership* 46:7 (1989), 41-47; Dennie P. Wolf et al., "To Use Their Minds Well: Investigating New Forms of Student Assessment," in Gerald Grand, ed., *Review of Research in Education* (Washington, D.C.: American Educational Research Association, 1991), 31-74; Daniel Resnick and Lauren Resnick, "Assessing the Thinking Curriculum: New Tools for Educational Reform," in Bernard R. Gifford and Mary C. O'Connor, eds., *Changing Assessments: Alternative Views of Aptitude, Achievement, and Instruction* (Boston, MA: Kluwer Academic Publishers, 1992), 37-75; U.S. Congress, Office of Technology Assessment, *Testing in American Schools: Asking the Right Questions,* OTA-SET-519 (Washington, D.C.: U.S. Government Printing Office, 1992).

[2] O. W. Caldwell and S. A. Courtis, *Then and Now in Education, 1845-1923: A Message of Encouragement from the Past to the Present* (Yonkers-on-Hudson, NY: World Book, 1925), 180-181.

[3] Harlan H. Horner, *Education in New York State, 1784-1954* (Albany, NY: University of the State of New York, State Education Department, 1954), 70.

[4] Herbert E. Hawkes, E. F. Lindquist, and C. R. Mann, *The Construction and Use of Achievement Examinations: A Manual for Secondary School Teachers* (Boston, MA: Houghton Mifflin, 1936), iv.

[5] George F. Madaus et al., *The Influence of Testing on Teaching Math and Science in Grades 4-12: Executive Summary* (Boston, MA: Boston College, Center for the Study of Testing, Evaluation, and Educational Policy, 1992), 1.

[6] *Testing in American Schools: Asking the Right Questions*, 64-65; National Council on Education Standards and Testing, *Raising Standards for American Education: A Report to Congress, the Secretary of Education, the National Education Goals Panel, and the American People* (Washington, D.C.: U.S. Government Printing Office, 1992), 12; See also *The Influence of Testing on Teaching Math and Science in Grades 4-12: Executive Summary,* 18.

[7] *Raising Standards for American Education*; Edward Silver, "Assessment and Mathematics Education Reform in the United States, *International Journal of Educational Research*, in press; Andrew C. Porter, "Assessing National Goals: Some Measurement Dilemmas," *The Assessment of National Educational Goals: Invitational Conference Proceedings* (Princeton, NJ: Educational Testing Service, 1970), 21-42; Lorrie Shepard, "Inflated Test Score Gains: Is the Problem Old Norms or Teaching to the Test?" *Educational Measurement: Issues and Practices* 9 (1990): 15-22; *Testing in American Schools: Asking the Right Questions.*

[8] Thomas Romberg, E. Anne Zarinnia, and Kevin F. Collis, "A New World View of Assessment in Mathematics," in Gerald Kulm, ed., *Assessing Higher*

Order Thinking in Mathematics (Washington, D.C.: American Association for the Advancement of Science, 1990), 21-38; Thomas Romberg, "Evaluation: A Coat of Many Colors" (Paper presented at the Sixth International Congress on Mathematical Education, Budapest, Hungary, July 27-August 3, 1988), Division of Science, Technical and Environmental Education, UNESCO; Edward Silver and Patricia Kenney, "Sources of Assessment Information for Instructional Guidance in Mathematics," in Thomas Romberg, ed., *Reform in School Mathematics and Authenic Assessment*, (in press); Edward Silver, Patricia Kenney, and Leslie Salmon-Cox, *The Content and Curricular Validity of the 1990 NAEP Mathematics Items: A Retrospective Analysis* (Pittsburgh, PA: Learning Research and Development Center, University of Pittsburgh, 1991); Richard Lesh and Susan J. Lamon, eds., *Assessment of Authentic Performance in School Mathematics* (Washington, D.C.: American Association for the Advancement of Science, 1992).

[9] *From Gatekeeper to Gateway*; Walter M. Haney, George F. Madaus, and Robert Lyons, *The Fractured Marketplace for Standardized Testing* (Boston, MA: Kluwer Academic Publishers), 319-323.

[10] Eva L. Baker and Harold F. O'Neil, Jr., "Diversity, Assessment, and Equity in Educational Reform" (Paper presented at the Ford Foundation Symposium on Equity and Educational Testing and Assessment, Washington, D.C., 11-12 March 1993).

[11] "Symbolic Validation"; Daniel M. Koretz et al., Statement before the Subcommittee on Elementary, Secondary, and Vocational Education, Committee on Education and Labor, U.S. House of Representatives (19 February 1992); Eva L. Baker, Harold F. O'Neill, and Robert L. Linn, "What Works in Performance Assessment" (Sherman Oaks, CA: Horace Design Information, Inc., Draft final report, September 1992); Eva L. Baker, "The Role of Domain Specifications in Improving the Technical Quality of Performance Assessment" (Los Angeles, CA: National Center for Research on Evaluation, Standards, and Student Testing, 1992); Norman Webb and E. Yasui, "Alternative Approaches to Assessment in Mathematics and Science: The Influence of Problem Context on Mathematics Performance" CSE Technical Report 346 (Los Angeles, CA: National Center for Research on Evaluation, Standards, and Student Testing, 1992).

[12] Stephen B. Dunbar and Elizabeth A. Witt, "Design Innovations in Measuring Mathematics Achievement" (Paper commissioned by the Mathematical Sciences Education Board, September 1993, appended to this report).

[13] For a discussion of relevant technical issues, *see* "Design Innovations in Measuring Mathematics Achievement" and others; Stephen B. Dunbar, Daniel M. Koretz, and H.D. Hoover, "Quality Control in the Development and Use of Performance Assessments," *Applied Measurement in Education* 4:4 (1991), 289-304; M. Magone et al., "Validity Evidence for Cognitive Complexity of Performance Assessments: An Analysis of Selected QUASAR Tasks" (Paper presented at the annual meeting of the American Educational Research Association, San Franciso, CA, April 1992); Daniel M. Koretz et al., "The Effects of High-Stakes Testing on Achievement: Preliminary Findings about Generalization Across Tests" (Paper presented at the annual meeting of the American Educational Research Association, Chicago, IL, April 1991); Robert Glaser, Kalyani Raghavan, and Gail P. Baxter, *Cognitive Theory as the Basis for Design of Innovative Assessment* (Los Angeles, CA: The Center for Research on Evaluation, Standards, and Student Testing, 1993).

3 ASSESSING IMPORTANT MATHEMATICAL CONTENT

High-quality mathematics assessment must be shaped and defined by important mathematical content. This fundamental concept is embodied in the first of three educational principles to guide assessment.

THE CONTENT PRINCIPLE
Assessment should reflect the mathematics that is most important for students to learn.

The content principle has profound implications for designing, developing, and scoring mathematics assessments as well as reporting their results. Some form of the content principle may have always implicitly guided assessment development, but in the past the notion of content has been construed in the narrow topic-coverage sense. Now content must be viewed much more broadly, incorporating the processes of mathematical thinking, the habits of mathematical problem solving, and an array of mathematical topics and applications, and this view must be made explicit. What follows is, nonetheless, a beginning description; much remains to be learned from research and from the wisdom of expert practice.

DESIGNING NEW ASSESSMENT FRAMEWORKS

Many of the assessments in use today, such as standardized achievement tests in mathematics, have reinforced the view that the mathematics curriculum is built from lists of narrow, isolated skills that can easily be decomposed for appraisal. The new vision of mathematics requires that assessment reinforce a new conceptualization that is both broader and more integrated.

The new vision

of mathematics

requires that

assessment

reinforce a new

conceptualization

that is both

broader and

more integrated.

Tests have traditionally been built from test blueprints, which have often been two dimensional arrays with topics to be covered along one axis and types of skills (or processes) on the other.[1] The assessment is then created by developing questions that fit into one cell or another of this matrix. But important mathematics is not always amenable to this cell-by-cell analysis.[2] Assessments need to involve more than one mathematical topic if students are to make appropriate connections among the mathematical ideas they have learned. Moreover, challenging assessments are usually open to a variety of approaches, typically using varied and multiple processes. Indeed, they can and often should be designed so that students are rewarded for finding alternative solutions. Designing tasks to fit a single topic and process distorts the kinds of assessments students should be able to do.

BEYOND TOPIC-BY-PROCESS FORMATS

Assessment developers need characterizations of the important mathematical knowledge to be assessed that reflect both the necessary coverage of content and the interconnectedness of topics and process. Interesting assessment tasks that do not elicit important mathematical thinking and problem solving are of no use. To avoid this, preliminary efforts have been made on several fronts to seek new ways to characterize the learning domain and the corresponding assessment. For example, lattice structures have recently been proposed as an improvement over matrix classifications of tasks.[3] Such structures provide a different and perhaps more interconnected view of mathematical understanding that should be reflected in assessment.

The approach taken by the National Assessment of Educational Progress (NAEP) to develop its assessments is an example of the effort to move beyond topic-by-process formats. Since its inception, NAEP has used a matrix design for developing its mathematics assessments. The dimensions of these designs have varied over the years, with a 35-cell design used in 1986 and the design below for the 1990 and 1992 assessments. Although classical test theory strongly encouraged the use of matrices to structure and provide balance to examinations, the designs also were often the root cause of the decontextualizing of assessments. If 35 percent of the items on the assessment were to be from the area of measurement and 40 percent of those were to assess students' procedural knowledge, then 14 percent of the items would measure procedural knowledge in the content domain of measurement. These items were developed to suit one cell of the matrix, without adequate consideration to the context and connections to other parts of mathematics.

Starting with the 1995 NAEP mathematics assessment, the use of matrices as a design feature has been discontinued. Percentages of items will be specified for each of the five major content areas, but some of these items will be double-coded because they measure content in more than one of the domains. Mathematical abilities categories—conceptual understanding, procedural knowl-

---

NAEP 1990-1992 Matrix

		Content				
		Numbers and Operations	Measurement	Geometry	Data Analysis, Probability, and Statistics	Algebra and Functions
Mathematical Ability	Conceptual Understanding					
	Procedural Knowledge					
	Problem Solving					

---

edge, and problem solving—will come into play only at the final stage of development to ensure that there is balance among the three categories over the entire assessment (although not necessarily by each content area) at each grade level. This change, along with the continued use of items requiring students to construct their own responses, has helped provide a new basis for the NAEP mathematics examination.[4]

One promising approach to assessment frameworks is being developed by the Balanced Assessment Project, which is a National Science Foundation-supported effort to create a set of assessment packages, at various grade levels, that provide students, teachers, and administrators with a fair and deep characterization of student attainment in mathematics.[5] The seven main dimensions of the framework are sketched below:

- content (which is very broadly construed to include concepts, senses, procedures and techniques, representations, and connections),

- thinking processes (conjecturing, organizing, explaining, proving, etc.),

- products (plans, models, reports, etc.),

- mathematical point of view (real-world modeling, for example),

- diversity (accessibility, sensitivity to language and culture, etc.),

- circumstances of performance (amount of time allowed, whether the task is to be done individually or in groups, etc.), and

- pedagogics-aesthetics (the extent to which a task or assessment is believable, engaging, etc.).

The first four dimensions describe aspects of the mathematical competency that the students are asked to demonstrate, whereas the last three dimensions pertain to characteristics of the assessment itself and the circumstances or conditions under which the assessment is undertaken.

One noteworthy feature of the framework from the Balanced Assessment Project is that it can be used at two different levels: at the level of the individual task and at the level of the assessment as a whole. When applied to an individual task, the framework can be used as more than a categorizing mechanism: it can be used to enrich or extend tasks by suggesting other thinking processes that might be involved, for example, or additional products that students might be asked to create. Just as important, the framework provides a way of examining the balance of a set of tasks that goes beyond checking off cells in a matrix. Any sufficiently rich task will involve aspects of several dimensions and hence will strengthen the overall balance of the entire assessment by contributing to several areas. Given a set of tasks, one can then examine the extent to which each aspect of the framework is represented, and this can be done without limiting oneself to tasks that fit entirely *inside* a particular cell in a matrix.

As these and other efforts demonstrate, researchers are attempting to take account of the fact that assessment should do much more than test discrete procedural skills.[6] The goal ought to be schemes for assessment that go beyond matrix classification to assessment that elicits student work on the meaning, process, and uses of mathematics. Although the goal is clearly defined, methods to achieve it are still being explored by researchers and practitioners alike.

SPECIFYING ASSESSMENT FRAMEWORKS

Assessment frameworks provide test developers with the guidance they need for creating new assessments. Embedded in the framework should be information to answer the following kinds of questions: What mathematics should students know before undertaking an assessment? What mathematics might they learn from the assessment? What might the assessment reveal about their understanding and their mathematical power? What mathematical background are they assumed to have? What information will they be given before, during, and after the assessment? How might the tasks be varied, extended, and incorporated into current instruction?

Developers also need criteria for determining appropriate student behavior on the assessment: Will students be expected to come up with conjectures on their own, for example, or will they be given some guidance, perhaps identification of a faulty conjecture, which can then be replaced by a better one? Will they be asked to write a convincing argument? Will they be expected to explain their conjecture to a colleague or to the teacher? What level of conjecture and argument will be deemed satisfactory for these tasks? A complete framework might also include standards for student performance (i.e., standards in harmony with the desired curriculum).

Very few examples of such assessment frameworks currently exist. Until there are more, educators are turning to curriculum frameworks, such as those developed by state departments of education

An assessment framework should provide a way to examine the balance of a set of tasks that goes beyond checking off cells in a matrix.

across the country, and adapting them for assessment purposes. The state of California, for example, has a curriculum framework that asserts the primacy of developing mathematical power for all students: "Mathematically powerful students think and communicate, drawing on mathematical ideas and using mathematical tools and techniques."[7] The framework portrays the content of mathematics in three ways:

- *Strands* (such as number, measurement, and geometry) run throughout the curriculum from kindergarten through grade 12. They describe the range of mathematics to be represented in the curriculum and provide a way to assess its balance.

- *Unifying ideas* (such as proportional relationships, patterns, and algorithms) are major mathematical ideas that cut across strands and grades. They represent central goals for learning and set priorities for study, bringing depth and connectedness, to the student's mathematical experience.

- *Units of instruction* (such as dealing with data, visualizing shapes, and measuring inaccessible distances) provide a means of organizing teaching. Strands are commingled in instruction, and unifying ideas give too big a picture to be useful day to day. Instruction is organized into coherent, manageable units consisting of investigations, problems, and other learning activities.

Through the California Learning Assessment System, researchers at the state department of education are working to create new forms of assessment and new assessment tasks to match the curriculum framework.[8]

Further exploration is needed to learn more about the development and appropriate use of assessment frameworks in mathematics education. Frameworks that depict the complexity of mathematics enhance assessment by providing teachers with better targets for teaching and by clearly communicating what is valued to students, their parents, and the general public.[9] Although an individual assessment may not treat all facets of the framework, the collection of assessments needed to evaluate what students are learning should be comprehensive. Such completeness is necessary if assessments are to provide the right kind of leadership for educational change. If an assessment represents a significant but small fraction of important mathematical knowledge and performance, then the same assessment should not be used over and over again. Repeated use could inappropriately narrow the curriculum.

DEVELOPING NEW ASSESSMENT TASKS

Several desired characteristics of assessment tasks can be deduced from the content principle and should guide the development of new assessment tasks.

TASKS REFLECTING MATHEMATICAL CONNECTIONS

Current mathematics education reform literature emphasizes the importance of the interconnections among mathematical topics and the connections of mathematics to other domains and disciplines. Much assessment tradition is based, however, on an atomistic approach that in practice, if not in theory, hides the connections among aspects of mathematics and between mathematics and other domains. Assessment developers will need to find new ways to reflect these connections in the assessment tasks posed for students.

One way to help ensure the interconnectedness is to create tasks that ask students to bring to bear a variety of aspects of mathematics. An example involving topics from arithmetic, geometry, and measurement appears on the following page.[10] Similarly, tasks may ask students to draw connections across various disciplines. Such tasks may provide some structure or hints for the students in finding the connections or may be more open-ended, leaving responsibility for finding connections to the students. Each strategy has its proper role in assessment, depending on the students' experience and accomplishment.

Another approach to reflecting important connections is to set tasks in a real-world context. Such tasks will more likely capture students' interest and enthusiasm and may also suggest new ways of understanding the world through mathematical models so that the assessment becomes part of the learning process. Moreover, the "situated cognition" literature[11] suggests that the specific settings and

One way to estimate the distance from where lightning strikes to you is to count the number of seconds until you hear the thunder and then divide by five. The number you get is the approximate distance in miles.

One person is standing at each of the four points A, B, C, and D. They saw lightning strike at E. Because sound travels more slowly than light, they did not hear the thunder right away.

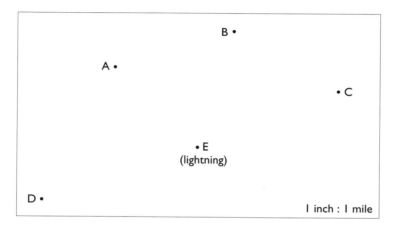

1. Who heard the thunder first? _____ Why?
   Who heard it last? _____ Why?
   Who heard it after 17 seconds? _____ Explain your answer.

2. How long did the person at B have to wait to hear the thunder? _____

3. Now suppose lightning strikes again at a *different* place. The person at A and the person at C both hear the thunder after the same amount of time. Show on the map below where the lightning might have struck.

B •

A •

• C

D •

1 inch : 1 mile

4. In question 3, are there other places where the lightning could have struck? _____
Explain your answer.

contexts in which a mathematical situation is embedded are critical determinants of problem solvers' responses to that situation. Developers should not assume, however, that just because a mathematical task is interesting to students, it therefore contains important mathematics. The mathematics in the task may be rather trivial and therefore inappropriate.

Test items that assess one isolated fragment of a student's mathematical knowledge may take very little time and may yield reliable scores when added together. However, because they are set in no reasonable context, they do not provide a full picture of the student's reasoning. They cannot show how the student connects mathematical ideas, and they seldom allow the student an opportunity to explain or justify a line of thinking.

Students should be clear about the context in which a question is being asked. Either the assumptions necessary for students to use mathematics in a problem situation should be made clear in the instructions or students should be given credit for correct reasoning under various assumptions. The context of a task, of course, need not be derived from mathematics. The example at right contains a task from a Kentucky state-wide assessment for twelfth-graders that is based on the notion of planning a budget within certain practical restrictions.[12]

Budget Planning Task

You graduated from Fairdale High School 2 years ago, and although you did not attend college, you have been attending night school to learn skills to repair video cassette recorders while you worked for minimum wages at a video center by day. Now you have been fortunate to find an excellent job that requires the special skills you have developed. Your salary will be $18,000.

This new job excites you because for some time you have been wanting to move out of your parents' home to your own apartment. During the past 2 years you have been able to buy your own bedroom set, a television, a stereo, and some of your own dishes and utensils.

To move to your own apartment, you will need to develop a budget. Your assignment is to develop a monthly budget showing how you will live on the income from your new job. To guide you, read the list below. (A packet of resource materials is provided, including a newspaper and brochures with consumer information.)

A. Estimate your monthly take-home pay. Remember that you must allow for city, state, federal, social security, and property taxes. Assume that city, state, federal, and social security taxes are 25% of your gross pay.

B. Using the newspaper provided, investigate various apartments and decide which one you will rent.

C. You will need a car on your new job. Price several cars and decide how much money you will need to borrow to buy the car you select; estimate the monthly payment. Use the newspaper and other consumer materials provided to make your estimate. Property taxes will be $10 per $1,000 assessed value.

D. You will do your own cooking. Figure how much you will spend on food, cooking and eating out.

E. As you plan your budget, don't forget about clothing, savings, entertainment and other living expenses.

Your budget for this project should be presented as a one-page, two-column display. Supporting this one-page budget summary, you should submit an explanation for each budget figure, telling how/where you got the information.

Other examples of age-appropriate contexts can be found in the fourth-grade assessments developed by the New Standards Project (NSP), a working partnership of researchers and state and local school districts formed to develop new systems of performance-based assessments. One such problem includes a fairly complex task in which children are given a table of information about various kinds of tropical fish (their lengths, habits, prices, etc.) and are asked to propose how to spend a fixed amount of money to buy a variety of fish for an aquarium of limited capacity, under certain realistic constraints.[13] The child must develop a solution that takes the various constraints into account. The task offers ample possibilities for students to display reasoning that connects mathematics with the underlying content.

THE CHALLENGES IN MAKING CONNECTIONS

The need to reflect mathematical connections pushes task development in new directions, each presenting challenges that require attention.

Differential Familiarity   Whatever the context of a mathematical task, some students will be more familiar with it than other students, possibly giving some an unfair advantage. One compensating approach is to spend time acquainting all students with the context. The NSP, for example, introduces the context of a problem in an assessment exercise in a separate lesson, taught before the assessment is administered.[14] Presumably the lesson reduces the variability among the students in their familiarity with the task setting. The same idea can be found in some of the assessment prototypes in *Measuring Up: Prototypes for Mathematics Assessment*. In one prototype, for instance, a script of a videotaped introduction was suggested;[15] playing such a videotape immediately before students work on the assessment task helps to ensure that everyone is equally familiar with the underlying context.

Another approach is to make the setting unusual, yet realistic, so that everyone will be starting with a minimum of prior knowledge. This technique was used in a study of children's problem solving conducted through extended individual task-based interviews.[16] The context used as the basis of the problem situation—a complex game involving probability—was deliberately constructed so that it would be unfamiliar to everyone. After extensive pilot testing of many variations,

Assessment

tasks can use

unusual, yet

realistic settings,

so that

everyone's prior

knowledge of

the setting is the

same.

an abstract version of the game was devised in which children's prior feelings and intuitive knowledge about winning and losing (and about competitions generally) could be kept separate from their mathematical analyses of the situation.

Clarifying Assumptions Task developers must consider seriously the impact of assumptions on any task, particularly as the assumptions affect the mathematics that is called for in solution of the problem. An example of the need to clarify assumptions is a performance assessment[17] that involves four tasks, all in the setting of an industrial arts class and all involving measuring and cutting wood. As written the tasks ignore an important idea from the realm of wood shop: When one cuts wood with a saw, a small but significant amount of wood is turned into sawdust. This narrow band of wood, called the saw's *kerf,* must always be taken into account, for otherwise the measurements will be off. The tasks contain many instances of this oversight: If, for example, a 16-inch piece is cut from a board that is 64 inches long, the remaining piece is not 48 inches long. Thus students who are fully familiar with the realities of wood shop could be at a disadvantage, since the problems posed are considerably more difficult when kerf is taken into account. Any scoring guide should provide an array of plausible answers for such tasks to ensure that students who answer the questions more accurately in real-world settings are given ample credit for their work. Better yet, the task should be designed so that assumptions about kerf (in this case) are immaterial to a solution.

Another assessment item[18] that has been widely discussed[19] also shows the need to clarify assumptions. In 1982, this item appeared in the third NAEP mathematics assessment: "An army bus holds 36 soldiers. If 1128 soldiers are being bussed to their training site, how many buses are needed?" The responses have been taken as evidence of U.S. students' weak understanding of mathematics, because only 33 percent of the 13-year-old students surveyed gave 32 as the answer, whereas 29 percent gave the quotient 31 with a remainder, and 18 percent gave just the quotient 31. There are of course many possible explanations as to why students who performed the division failed to give the expected whole-number answer. One plausible explanation may be that some students did not see a need to use one more bus to transport the remaining 12 soldiers. They could squeeze into the other buses; they could go by car. Asked about their answers in interviews or in writing, some

Task developers must consider whether students' assumptions affect the mathematics called for in solution of a problem.

students offer such explanations.[20] The point is that the answer 32 assumes that no bus can hold more than 36 soldiers and that alternative modes of transportation cannot be arranged.

Ease of Entry and Various Points of Exit   Students should be allowed to become engaged in assessment tasks through a sequence of questions of increasing difficulty. They ought to be able to exit the task at different points reflecting differing levels of understanding and insight into the mathematical content. For too long, assessment tasks have not provided all students with the opportunity to start the task, let alone work part way through it. As a result some students have inadequate opportunities to display their understanding. The emerging emphasis on connections lets assessment tasks be designed to permit various points of entry and exit.

As an example consider a problem from the California Assessment Program's Survey of Academic Skills.[21] The task starts with a square, into which a nested sequence of smaller squares is to be drawn, with the vertices of each square connecting the midpoints of the sides of the previous one. Starting with this purely geometric drawing task, the student determines the sequence of numbers corresponding to the areas of the squares and then writes a general rule that can be used to find the area of the nth interior square. Such a task allows the student who may not be able to produce a general rule at least to demonstrate his or her understanding of the geometrical aspects of the situation.

TASKS REQUIRING COMMUNICATION

The vision of mathematics and mathematics assessment described in earlier chapters emphasizes communication as an critical feature. Developers are beginning to recognize that there are many ways to communicate about mathematical ideas and that assessment tasks have seldom made sufficient use of these alternatives.

Incorporating communication about mathematics into assessment tasks obviously calls for different forms of responses than have been common in the past. Students may respond in a wide variety of ways: by writing an essay, giving an oral report, participating in a group discussion, constructing a chart or graph, or programming a computer.

Few current assessment tasks provide all students with the opportunity to start the task, let alone work part way through it.

In the aquarium task cited above, students were asked to write a letter to their principal explaining what fish they would buy and why they made their choices. In a task from an Educational Testing Service program for middle school mathematics,[22] students are asked to predict future running speed for male and female athletes, typically derived from graphs of tables they have constructed, and to justify their predictions in written form.

Some assessments can be carried out as individual interviews. The example below describes a question on an oral examination for 17-year-old Danish mathematics students in the Gymnasium.[23] (The Gymnasium enrolls less than 30 percent of the age cohort, and not all Gymnasium students take mathematics.) One clear benefit of an oral assessment is that it allows the assessor to identify immediately how students are interpreting the problem context and what assumptions they are making.

Tasks that require students to communicate about mathematics pose the following challenge: To what extent are differences in ability to communicate to be considered legitimate differences in mathematical power? Clearly efforts must be made to ensure that students are given the opportunity to respond to assessments in the language they speak, read, and write best. Different students will choose different preferred modes of mathematical communication. For example, some will be able to explain their reasoning more effectively with a table or chart than with an equation or formula; for others the reverse will be true. Hence tasks should be constructed, insofar as possible, to allow various approaches and various

---

Oral Assessments

Expound on the exponential growth model, including formulas and graphs (the following data may be used as a basis).

Under favorable circumstances the bacterium *Escherichia coli* divides every 20 minutes:

Time (min)	0	20	40	60	80	...
Number	1000	2000	4000	8000	16000	...

The hour wages (in Danish kroners) of female workers in Denmark were for the years 1963-1970:

Year	1963	1964	1965	...
Hourly wage	5.97	6.61	7.43	...

At the oral examination it is expected that the student, unassisted, will explain, e.g.,

- What sort of growth in the real world may be exponential and why?
- Why does $f(x) = ba^x$ describe constant growth in percentage?
- Why is $f(x) = ba^x$ equivalent to the graph of f being a straight line in a semi-logarithmic coordinate system?

To the
extent that
communication
is a part of
mathematics,
differences in
communication
skill must
be seen as
differences in
mathematical
power.

response modes. The scoring of the assessments must take the variety of valid approaches into account.

Nonetheless, some differences in performance will remain. To the extent that communication is a part of mathematics, differences in communication skill must be seen as differences in mathematical power. Means of fairly evaluating responses, accounting for both the student's and the assessor's preferred modes of communication for any given task, must be developed.

Of course, communication is a two-way street. Questions should be understandable and clear. The need for careful use of language and notation in stating a task has long been a goal of assessment developers, although one not always successfully achieved.

The example below from the Second International Mathematics Study (SIMS) illustrates this difficulty in a multiple-choice item.[24] The aim of the item is evidently to tap students' abilities to apply the Pythagorean Theorem to a right triangle formed at the top of the figure by drawing a segment parallel to the bottom segment, concluding that x equals 8 m. Because the question is posed in a multiple-choice format, however, and because the figure is drawn so close to scale, the correct choice, C, can be found by estimating the length visually. Certainly this is a fully legitimate, and indeed more

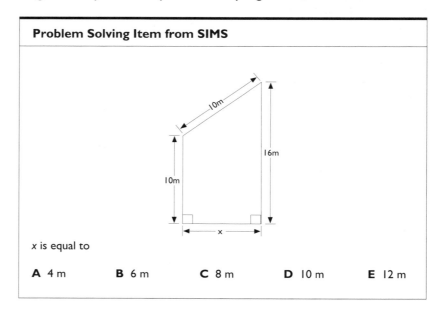

Problem Solving Item from SIMS

x is equal to

A 4 m   **B** 6 m   **C** 8 m   **D** 10 m   **E** 12 m elegant, way of solving the problem than the posers had probably intended, but if the goal is to determine students' knowledge of the Pythagorean relationship, then the item is misguided. Figures, like all other components of a task, must be carefully examined to see if they convey what the assessment designer intends.

SOLVING NONROUTINE PROBLEMS

Problem solving is the cornerstone of the reform effort in mathematics education. The types of problems that matter—the types we really wish to have students learn how to solve—are the ones that are nonroutine. It is not sufficient for students to be able to solve highly structured, even formulaic problems that all require the same approach. Students must face problems they have not encountered before and learn how to approach and solve them. Thus, in an assessment, to learn what a student can do with nonroutine problems, the problems must be genuinely new to the student. The goal is to draw on the habits of thinking developed by students in their studies, not on specific problems they have learned to solve in their earlier work.

There is some tension between the claim that nonroutine problems are the most legitimate kind of problem, on the one hand, and the need for fairness, on the other. Tasks must be constructed in such a way that the intended audience has all the prerequisite skills (or should have had all the prerequisite skills) yet has never seen a problem just like the one it is confronted with on the assessment. Nonroutine problems pose special issues for the security of assessment. To be novel, assessment tasks must not have been seen in advance. However, students should be exposed to a variety of novel problems in daily work to demonstrate to them that nonroutine problems are valued in mathematics education and that they should expect to see many such problems.

The challenge, ultimately, is to ensure that all students being assessed have had substantial experience in grappling with nonroutine problems as well as the opportunity to learn the mathematical ideas embedded in the problem. Teachers must be open to alternative solutions to nonroutine problems. When instruction has not been directed to preparing students for nonroutine problem solving, performance will likely be related more to what has been called aptitude than to instructionally related learning.

The goal of assessment is to draw on the habits of thinking developed by students in their studies, not on specific problems they have learned to solve in their earlier work.

New kinds of

assessments call

for new kinds of

expertise among

those who

develop the

tasks.

MATHEMATICAL EXPERTISE

New kinds of assessments call for new kinds of expertise among those who develop the tasks. The special features of the mathematics content and the special challenges faced in constructing assessment tasks illustrate a need for additional types of expertise in developing assessment tasks and evaluation schema. Task developers need to have a high level of understanding of children, how they think about things mathematical and how they learn mathematics, well beyond the levels assumed to be required to develop assessment tasks in the past. Developers must also have a deep understanding of mathematics and its applications. We can no longer rely on task developers with superficial understanding of mathematics to develop assessment tasks that will elicit creative and novel mathematical thinking.

SCORING NEW ASSESSMENTS

The content principle also has implications for the mathematical expertise of those who score assessments and the scoring approaches that they use.

JOINING TASK DEVELOPMENT TO STUDENT RESPONSES

A multiple-choice question is developed with identification of the correct answer. Similarly, an open-ended task is incomplete without a scoring rubric—a scoring guide—as to how the response will be evaluated. Joining the two processes is critical because the basis on which the response will be evaluated has many implications for the way the task is designed, and the way the task is designed has implications for its evaluation.

Just as there is a need to try out multiple-choice test questions prior to administration, so there is a need to try out the combination of task and its scoring rubric for open-ended questions. Students' responses give information about the design of both the task and the rubric. Feedback loops, where assessment tasks are modified and sharpened in response to student work, are especially important, in part because of the variety of possible responses.

EVALUATING RESPONSES TO REFLECT THE CONTENT PRINCIPLE

The key to evaluating responses to new kinds of assessment tasks is having a scoring rubric that is tied to the prevailing vision of mathematics education. If an assessment consists of multiple-choice items, the job of determining which responses are correct is straightforward, although assessment designers have little information to go on in trying to decide *why* students have made certain choices. They can interview students after a pilot administration of the test to try to understand why they chose the answers they did. The designers can then revise the item so that the erroneous choices may be more interpretable. If ambiguity remains and students approach the item with sound interpretations that differ from those of the designers, the response evaluation cannot help matters much. The item is almost always scored either right or wrong.[25]

Designers of open-ended tasks, on the other hand, ordinarily describe the kinds of responses expected in a more general way. Unanticipated responses can be dealt with by judges who discuss how those responses fit into the scoring scheme. The standard-setting process used to train judges to evaluate open-ended responses, including portfolios, in the Advanced Placement (AP) program of the College Board, for example, alternates between the verbal rubrics laid out in advance and samples of student work from the assessment itself.[26] Portfolios in the AP Studio Art evaluation are graded by judges who first hold a standard-setting session at which sample portfolios representing all the possible scores are examined and discussed. The samples are used during the judging of the remaining portfolios as references for the readers to use in place of a general scoring rubric. Multiple readings and moderation by more experienced graders help to hold the scores to the agreed standard.[27] Together, graders create a shared understanding of the rubrics they are to use on the students' work. Examination boards in Britain follow a similar procedure in marking students' examination papers in subjects such as mathematics, except that a rubric is used along with sample examinations discussed by the group to help examiners agree on marks.[28]

The development of high-quality scoring guides to match new assessment is a fairly recent undertaking. One approach has been first to identify in general terms the levels of desired performance and then to create task-specific rubrics. An example from a New Jersey eighth-grade "Early Warning" assessment appears on the following page.[29]

A general rubric can be used to support a holistic scoring system, as New Jersey has done, in which the student's response is examined and scored as a whole. Alternatively, a much more refined analytic scheme could be devised in which specific features or qualities of a student's response are identified, according to predetermined criteria, and given separate scores. In the example from New Jersey, one can imagine a rubric that yields two independent scores: one for the accuracy of the numerical answer and one for the adequacy of the explanation.

Assessors are experimenting with both analytic and holistic approaches, as well as a amalgam of the two. For example, in the Mathematics Performance Assessment developed by The Psychological Corporation,[30] responses are scored along the dimensions of reasoning, conceptual knowledge, communication, and procedures, with a separate rubric for each dimension. In contrast, QUASAR, a project to improve the mathematics instruction of middle school students in economically disadvantaged communities,[31] uses an approach that blends task-specific rubrics with a more general rubric, resulting in scoring in which mathematical knowledge, strategic knowledge, and communication are considered interrelated components. These components are not rated separately but rather are to be considered in arriving at a holistic rating.[32] Another approach is through so-called protorubrics, which were developed for the tasks in *Measuring Up*.[33] The protorubrics can be adapted for either holistic or analytic approaches and are designed to give only selected characteristics and examples of high, medium, and low responses.

Profound challenges confront the developer of a rating scheme regardless of the system of scoring or the type of rubric used. If a rubric is developed to deal with a single task or a type of task, the important mathematical ideas and processes involved in the task can be specified so that the student can be judged on how well those appear to have been mastered, perhaps sacrificing some degree of interconnectedness among tasks. On the other hand, general rubrics may not allow scorers to capture some important qualities of students' thinking about a particular task. Instead,

Profound

challenges

confront the

developer of a

rating scheme

regardless of

the system of

scoring or the

type of rubric

used.

From a Generalized Holistic Scoring Guide to a Specific Annotated Item Scoring Guide

Generalized Scoring Guide:

Student demonstrates proficiency — Score Point = 3.

The student provides a satisfactory response with explanations that are plausible, reasonably clear, and reasonably correct, e.g., includes appropriate diagram(s), uses appropriate symbols or language to communicate effectively, exhibits an understand of the mathematics of the problem, uses appropriate processes and/or descriptions to answer the question, and presents sensible supporting arguments. Any flaws in the response are minor.

Student demonstrates minimal proficiency — Score Point = 2

The student provides a nearly satisfactory response which contains some flaws, e.g., begins to answer the question correctly but fails to answer all of its parts or omits appropriate explanation, draws diagram(s) with minor flaws, makes some errors in computation, misuses mathematical language, or uses inappropriate strategies to answer the question.

Student demonstrates a lack of proficiency — Score Point = 1

The student provides a less than satisfactory response that only begins to answer the question, but fails to answer it completely, e.g., provides little or no appropriate explanation, draws diagram(s) which are unclear, exhibits little or no understanding of the question being asked, or makes major computational errors.

Student demonstrates no proficiency — Score Point = 0

The student provides an unsatisfactory response that answers the question inappropriately, e.g., uses algorithms which do not reflect any understanding of the question, makes drawings which are inappropriate to the question, provides a copy of the question without an appropriate answer, fails to provide any information which is appropriate to the question, or fails to attempt to answer the question.

Specific Problem:

What digit is in the fiftieth decimal place of the decimal form of 3/11? Explain your answer.

Annotated Scoring Guide:

3 points The student provides a satisfactory response; e.g., indicates that the digit in the fiftieth place is 7 and shows that the digits 2 and 7 in the quotient (.272727...) alternate; the explanation of why 7 is the digit in the fiftieth place is either based on some counting procedure or on the pattern of how the digits are positioned after the decimal point. (The student could read fiftieth as fifteenth or fifth, identify 2 as the digit, and provide an explanation similar to the ones above.)

2 points The student provides a nearly satisfactory response which contains some flaws, e.g., identifies the pattern of the digits 2 and 7 (.272727...) and provide either a weak or no explanation of why 7 is the digit in the fiftieth place OR converts 3/11 incorrectly to 3.666... and provides some explanation of why 6 is the digit in the fiftieth place.

1 point The student provides a less than satisfactory response that only begins to answer the question; e.g., begins to divide correctly (minor flaws in division are allowed) but fails to identify "the digit" OR identifies 7 as the correct digit with no explanation or work shown.

0 points The student provides an unsatisfactory response; e.g., either answers the question inappropriately or fails to attempt to answer the question.

anecdotal evidence suggests that students may be given credit for verbal fluency or for elegance of presentation rather than mathematical acumen. The student who mentions everything possible about the problem posed in the task and rambles on about minor points the teacher has mentioned in class may receive more credit than a student who has deeper insights into the problem but produces only a terse, minimalist solution. The beautiful but prosaic presentation with elaborate drawings may inappropriately outweigh the unexpected but elegant solution. Such difficulties are bound to arise when communication with others is emphasized as part of mathematical thinking, but they can be dealt with more successfully when assessors include those with expertise in mathematics.

Unanticipated responses require knowledgeable graders who can recognize and evaluate them.

In any case, regardless of the type of rubric, graders must be alert to the unconventional, unexpected answer, which, in fact, may contain insights that the assessor had not anticipated. The likelihood of unanticipated responses will depend in part upon the mathematical richness and complexity of the task. Of course, the greater the chances of unanticipated responses, the greater the mathematical sophistication needed by the persons grading the tasks: the graders must be sufficiently knowledgeable to recognize kernels of mathematical insight when they occur. Similarly, graders must sharpen their listening skills for those instances in which task results are communicated orally. Teachers are uniquely positioned to interpret their students' work on internal and external assessments. Personal knowledge of the students enhances their ability to be good listeners and to recognize the direction of their students' thinking.

There may also be a need for somewhat different rubrics even on the same task because judgment of draft work should be different from judgment of polished work. With problem solving a main thrust of mathematics education, there is a place for both kinds of judgments. Some efforts are under way, for example, to establish iterative processes of assessment: Students work on tasks, handing it in to teachers to receive comments about their work in progress. With these comments in hand, students may revise and extend their work. Again, it goes to the teacher for comment. This back-and-forth process may continue several times, optimizing the opportunity for students to learn from the assessment. Such a model will require appropriate rubrics for teachers and students alike to judge progress at different points.

Reporting Assessment Results

Consideration of issues about the dissemination of results are often not confronted until after an assessment has been administered. This represents a missed opportunity, particularly from the perspective of the content principle. Serious attention to what kind of information is needed from the assessment and who needs it should influence the design of the assessment and can help prevent some of the common misuses of assessment data by educators, researchers, and the public. The reporting framework itself must relate to the mathematics content that is important for all students to learn.

There has been a long tradition in external assessment of providing a single overall summary score, coupled in some cases with subscores that provide a more fine-grained analysis. The most typical basis for a summary score has been a student's relative standing among his or her group of peers. There have been numerous efforts to move to other information in a summary score, such as percent mastery in the criterion-related measurement framework. One innovative approach has been taken by the Western Australia Monitoring Standards in Education program. For each of five strands (number; measurement; space; chance and data; algebra) a student's performances on perhaps 20 assessment tasks are arrayed in such a way that overall achievement is readily apparent while at the same time some detailed diagnostic information is conveyed.[34] NAEP developed an alternative approach to try to give meaning to summary scores beyond relative standing. NAEP used statistical techniques to put all mathematics items in the same mathematics proficiency scale so that sets of items can be used to describe the level of proficiency a particular score represents.[35] Although these scales have been criticized for yielding misinterpretations about what students know and can do in mathematics,[36] they represent one attempt to make score information more meaningful.

Similarly, some teachers focus only on the correctness of the final answer on teacher-made tests with insufficient attention to the mathematical problem solving that preceded it. Implementation of the content principle supports a reexamination of this approach. Problem solving legitimately may involve some false starts or blind alleys; students whose work includes such things are doing important mathematics.

Rather than

forcing

mathematics to

fit assessment,

assessment

must be tailored

to whatever

mathematics is

important to

learn.

Along with the efforts to develop national standards in various fields, there is a push to provide assessment information in ways that relate to progress toward those national standards. Precisely how such scores would be designed to relate to national standards and what they would actually mean are unanswered questions. Nonetheless, this push also is toward reporting methods that tell people directly about the important mathematics students have learned. This is the approach that NAEP takes when it illustrates what basic, proficient, and advanced mean by giving specific examples of tasks at these levels.

An assessment framework that is used as the foundation for the development of an assessment may provide, at least in part, a lead to how results of the assessment might be reported. In particular, the major themes or components of a framework will give some guidance with regard to the appropriate categories for reporting. For example, the first four dimensions of the Balanced Assessment Project's framework suggest that attention be paid to describing students' performance in terms of thinking processes used and products produced as well as in terms of the various components of content. In any case, whether or not a direct connection between aspects of the framework and reporting categories is made, a determination of reporting categories should affect and be affected by the categories of an assessment framework.

The mathematics in an assessment should never be distorted or trivialized for the convenience of assessment. Design, development, scoring, and reporting of assessments must take into account the mathematics that is important for students to learn.

In summary, rather than forcing mathematics to fit assessment, assessment must be tailored to whatever mathematics is important to assess.

ENDNOTES

[1] For examples of such matrices, see Edward G. Begle and James W. Wilson, "Evaluation of Mathematics Programs," in Edward G. Begle, ed., *Mathematics Education*, 69th Yearbook of the National Society for the Study of Education, pt. 1 (Chicago, IL: University of Chicago Press, 1970), 367-404; for a critique of this approach to content, see Thomas A. Romberg, E. Anne Zarinnia, and Kevin F. Collis, "A New World View of Assessment in Mathematics," in Gerald Kulm, ed., *Assessing Higher Order Thinking in Mathematics* (Washington, D.C.: American Association for the Advancement of Science, 1990), 24-27.

[2] Edward A. Silver, Patricia Ann Kenney, and Leslie Salmon-Cox, *The Content and Curricular Validity of the 1990 NAEP Mathematics Items: A Retrospective Analysis* (Pittsburgh, PA: Learning Research and Development Center, University of Pittsburgh, 1991).

[3] Edward Haertel and David E. Wiley, "Representations of Ability Structures: Implications for Testing," in Norman Fredericksen, Robert J. Mislevy, and Isaac I. Bejar, eds., *Test Theory for a New Generation of Tests* (Hillsdale, NJ: Lawrence Erlbaum Associates, 1992).

[4] John Dossey, personal communication, 24 June 1993.

[5] Alan H. Schoenfeld, *Balanced Assessment for the Mathematics Curriculum: Progress Report to the National Science Foundation* (Berkeley, CA: University of California, June 1993).

[6] See also Suzanne P. Lajoie, "A Framework for Authentic Assessment in Mathematics," in Thomas A. Romberg, ed., *Reform in School Mathematics and Authentic Assessment,* in press.

[7] California Department of Education, *Mathematics Framework for California Public Schools: Kindergarten Through Grade 12* (Sacramento, CA: Author, 1992), 20.

[8] E. Anne Zarinnia and Thomas A. Romberg, "A Framework for the California Assessment Program to Report Students' Achievement in Mathematics," in Thomas A. Romberg, ed., *Mathematics Assessment and Evaluation: Imperatives for Mathematics Education* (Albany, NY: State University of New York Press, 1992), 242-284.

[9] Lauren B. Resnick and Daniel P. Resnick, "Assessing the Thinking Curriculum: New Tools for Educational Reform," in Bernard R. Gifford and Mary Catherine O'Connor, eds., *Changing Assessments: Alternative Views of Aptitude, Achievement and Instruction* (Boston, MA: Kluwer Academic Publishers, 1992), 37-75.

[10] National Research Council, Mathematical Sciences Education Board, *Measuring Up: Prototypes for Mathematics Assessment* (Washington, D.C.: National Academy Press, 1993), 117-119.

[11] John S. Brown, Allan Collins, and P. Duguid, "Situated Cognition and the Culture of Learning," *Educational Researcher* 18:1 (1989), 32-42; Ralph T. Putnam, Magdalene Lampert, and Penelope Peterson, "Alternative Perspectives on Knowing Mathematics in Elementary Schools," *Review of Research in Education* 16 (1990):57-150; James G. Greeno, "A Perspective on Thinking," *American Psychologist* 44:2 (1989), 134-141; James Hiebert and Thomas P. Carpenter, "Learning and Teaching with Understanding," in Douglas A. Grouws, ed., *Handbook of Research on Mathematics Teaching and Learning* (New York, NY: Macmillan Publishing Company, 1992), 65-97.

[12] Kentucky Department of Education, "All About Assessment," *EdNews*, Special Section, Jan/Feb 1992, 7.

[13] Lauren B. Resnick, Diane Briars, and Sharon Lesgold, "Certifying Accomplishments in Mathematics: The New Standards Examining System," in Izaak Wirszup and Robert Streit, eds., *Developments in School Mathematics Education Around the World*, vol. 3 (Reston, VA: National Council of Teachers of Mathematics, 1992), 196-200.

[14] Learning Research and Development Center, University of Pittsburgh and National Center on Education and the Economy, *New Standards Project* (Pittsburgh, PA: Author, 1993).

[15] *Measuring Up*, 101-106.

[16] Eve R. Hall, Edward T. Esty, and Shalom M. Fisch, "Television and Children's Problem-Solving Behavior: A Synopsis of an Evaluation of the Effects of *Square One TV*," *Journal of Mathematical Behavior* 9:2 (1990), 161-174.

[17] The Riverside Publishing Company, *Riverside Student Performance Assessment, Grade 8 Mathematics Sample Assessment* (Riverside, CA: Author, 1991), 2-6.

[18] Thomas P. Carpenter et al., "Results of the Third NAEP Mathematics Assessment: Secondary School," *Mathematics Teacher*, 76:9 (1983), 656.

[19] See, for example, Alan H. Schoenfeld, "When Good Teaching Leads to Bad Results: The Disasters of 'Well Taught' Mathematics Classes," *Educational Psychologist* 23:2 (1988), 145-166; Mary M. Lindquist, "Reflections on the Mathematics Assessments of the National Assessment of Educational Progress," in *Developments in School Mathematics Education Around the World*, vol. 3.

[20] Edward A. Silver, Lora J. Shapiro, and Adam Deutsch, "Sense Making and the Solution of Division Problems Involving Remainders: An Examination of Middle School Students' Solution Processes and Their Interpretations of Solutions," *Journal for Research in Mathematics Education* 24:2 (1993), 117-135.

[21] California Assessment Program, *Question E* (Sacramento, CA: California State Department of Education, 1987).

[22] Nancy S. Cole, *Changing Assessment Practice in Mathematics Education: Reclaiming Assessment for Teaching and Learning* (Draft version, 1992).

[23] Adapted from Kirsten Hermann and Bent Hirsberg, "Assessment in Upper Secondary Mathematics in Denmark," in Mogens Niss, ed., *Cases of Assessment in Mathematics Education* (Dordrecht, The Netherlands: Kluwer Academic Publishers, 1993), 133.

[24] Second International Mathematics Study, "Technical Report 4: Instrument Book," booklet 2LB, problem 26, (Urbana, IL: International Association for the Evaluation of Educational Achievement, 8, November 1985), 8. This was the only item, of those given to eighth graders in the U.S., that was later judged to have involved problem solving as specified in the NCTM *Standards*.

[25] Peter Hilton, "The Tyranny of Tests," *American Mathematical Monthly* 100:4 (1993), 365-369.

[26] "Representations of Ability Stuctures"; Robert J. Mislevy, "Test Theory Reconceived," Research Report, in press.

[27] Ruth Mitchell, *Testing for Learning: How New Approaches to Evaluation Can Improve American Schools* (New York: Free Press, 1992).

[28] Alan Bell, Hugh Burkhardt, and Malcolm Swan, "Assessment of Extended Tasks," in Richard Lesh and Susan J. Lamon, eds., *Assessment of Authentic Performance in School Mathematics* (Washington, D.C.: American Association for the Advancement of Science, 1992), 182.

[29] New Jersey Department of Education, "Grade 8 Early Warning Test," *Guide to Procedures for Scoring the Mathematics Constructed-Response Items* (Trenton, NJ: Author, 1991), 4-6.

[30] Marilyn Rindfuss, ed., *Integrated Assessment System: Mathematics Performance Assessment Tasks Scoring Guides* (San Antonio, TX: The Psychological Corporation, 1991).

[31] Edward A. Silver, "QUASAR," *Ford Foundation Letter*, 20:3 (1989), 1-3.

[32] Edward A. Silver and Suzanne Lane, "Assessment in the Context of Mathematics Instruction Reform: The Design of Assessment in the QUASAR Project," in *Cases of Assessment in Mathematics Education: An ICMI Study*.

[33] *Measuring Up*, 14-16.

[34] Geoff N. Masters, *Inferring Levels of Achievement on Profile Strands* (Hawthorn, Australia: Australian Council for Educational Research, 1993).

[35] John A. Dossey et al., *The Mathematical Report Card: Are We Measuring Up?* (Princeton, NJ: Educational Testing Service, 1988).

[36] Robert A. Forsyth, "Do NAEP Scales Yield Valid Criterion-Referenced Interpretations?" *Educational Measurement: Issues and Practice* 10:3 (1991), 3-9, 16. For a more recent critique of the procedures that the National Assessment Governing Board has used in setting and interpreting performance standards in the 1992 mathematics NAEP, see *Educational Achievement Standards: NAGB's Approach Yields Misleading Interpretations* (Washington, D.C.: General Accounting Office, 1993).

4 | ASSESSING TO SUPPORT MATHEMATICS LEARNING

igh-quality mathematics assessment must focus on the interaction of assessment with learning and teaching. This fundamental concept is embodied in the second educational principle of mathematics assessment.

THE LEARNING PRINCIPLE
Assessment should enhance mathematics learning and support good instructional practice.

This principle has important implications for the nature of assessment. Primary among them is that assessment should be seen as an integral part of teaching and learning rather than as the culmination of the process.[1] As an integral part, assessment provides an opportunity for teachers and students alike to identify areas of understanding and misunderstanding. With this knowledge, students and teachers can build on the understanding and seek to transform misunderstanding into significant learning. Time spent on assessment will then contribute to the goal of improving the mathematics learning of all students.

The applicability of the learning principle to assessments created and used by teachers and others directly involved in classrooms is relatively straightforward. Less obvious is the applicability of the principle to assessments created and imposed by parties outside the classroom. Tradition has allowed and even encouraged some assessments to serve accountability or monitoring purposes without sufficient regard for their impact on student learning.

A portion of assessment in schools today is mandated by external authorities and is for the general purpose of accountability of the schools. In 1990, 46 states had mandated testing programs, as compared with 20 in 1980.[2] Such assessments have usually been multiple-choice norm-referenced tests. Several researchers have studied these testing programs and judged them to be inconsistent with the current goals of mathematics education.[3] Making mandated assessments consonant with the content, learning, and equity principles will require much effort.

Instruction and

assessment—

from whatever

source and for

whatever

purpose—must

support one

another.

Studies have documented a further complication as teachers are caught between the conflicting demands of mandated testing programs and instructional practices they consider more appropriate. Some have resorted to "double-entry" lessons in which they supplement regular course instruction with efforts to teach the objectives required by the mandated test.[4] During a period of change there will undoubtedly be awkward and difficult examples of discontinuities between newer and older directions and procedures. Instructional practices may move ahead of assessment practices in some situations, whereas in other situations assessment practices could outpace instruction. Neither situation is desirable although both will almost surely occur. However, still worse than such periods of conflict would be to continue either old instructional forms or old assessment forms in the name of synchrony, thus stalling movement of either toward improving important mathematics learning.

From the perspective of the learning principle, the question of who mandated the assessment and for what purpose is not the primary issue. Instruction and assessment—from whatever source and for whatever purpose—must be integrated so that they support one another.

To satisfy the learning principle, assessment must change in ways consonant with the current changes in teaching, learning, and curriculum. In the past, student learning was often viewed as a passive process whereby students remembered what teachers told them to remember. Consistent with this view, assessment was often thought of as the end of learning. The student was assessed on something taught previously to see if he or she remembered it. Similarly, the mathematics curriculum was seen as a fragmented collection of information given meaning by the teacher.

This view led to assessment that reinforced memorization as a principal learning strategy. As a result, students had scant opportunity to bring their intuitive knowledge to bear on new concepts and tended to memorize rules rather than understand symbols and procedures.[5] This passive view of learning is not appropriate for the mathematics students need to master today. To develop mathematical competence, students must be involved in a dynamic process of thinking mathematically, creating and exploring methods of solution, solving problems, communicating their understanding— not simply remembering things. Assessment, therefore, must reflect and reinforce this view of the learning process.

This chapter examines three ways of making assessment compatible with the learning principle: ensuring that assessment directly supports student learning; ensuring that assessment is consonant with good instructional practice; and enabling teachers to become better facilitators of student learning.

ASSESSMENT IN SUPPORT OF LEARNING

Assessment can play a key role in exemplifying the new types of mathematics learning students must achieve. Assessments indicate to students what they should learn. They specify and give concrete meaning to valued learning goals. If students need to learn to perform mathematical operations, they should be assessed on mathematical operations. If they should learn to use those mathematical operations along with mathematical reasoning in solving mathematical problems, they must be assessed on using mathematical operations along with reasoning to solve mathematical problems. In this way the nature of the assessments themselves make the goals for mathematics learning real to students, teachers, parents, and the public.

Mathematics assessments can help both students and teachers improve the work the students are doing in mathematics. Students need to learn to monitor and evaluate their progress. When students are encouraged to assess their own learning, they become more aware of what they know, how they learn, and what resources they are using when they do mathematics. "Conscious knowledge about the resources available to them and the ability to engage in self-monitoring and self-regulation are important characteristics of self-assessment that successful learners use to promote ownership of learning and independence of thought."[6]

> Mathematics assessments can make the goals for learning real to students, teachers, parents, and the public.

In the emerging view of mathematics education, students make their own mathematics learning individually meaningful. Important mathematics is not limited to specific facts and skills students can be trained to remember but rather involves the intellectual structures and processes students develop as they engage in activities they have endowed with meaning.

> The assessment challenge we face is to give up old assessment methods to determine what students know, which are based on behavioral theories of learning and develop authentic assessment procedures that reflect current epistemological beliefs both about what it means to know mathematics and how students come to know.[7]

> Current research indicates that acquired knowledge is not simply a collection of concepts and procedural skills filed in long-term memory. Rather the knowledge is structured by individuals in meaningful ways, which grow and change over time.[8]

> A close consideration of recent research on mathematical cognition suggests that in mathematics, as in reading, successful learners understand the task to be one of constructing meaning, of doing interpretive work rather than routine manipulations. In mathematics the problem of imposing meaning takes a special form: making sense of formal symbols and rules that are often taught as if they were arbitrary conventions rather than expressions of fundamental regularities and relationships among quantities and physical entities.[9]

LEARNING FROM ASSESSMENT

Modern learning theory and experience with new forms of assessment suggest several characteristics assessments should have if they are to serve effectively as learning activities. Of particular interest is the need to provide opportunities for students to construct their own mathematical knowledge and the need to determine where students are in their acquisition of mathematical understanding.[10] One focuses more on the content of mathematics, the other on the process of doing mathematics. In both, the assessment must elicit important mathematics.

Constructing Mathematical Knowledge Learning is a process of continually restructuring one's prior knowledge, not just adding to it. Good education provides opportunities for students to connect what is being learned to their prior knowledge. One knows mathematics best if one has developed the structures and meanings of the content for oneself.[11] For assessment to support learning, it must enable students to construct new knowledge from what they know.

One way to provide opportunities for the construction of mathematical knowledge is through assessment tasks that resemble learning tasks[12] in that they promote strategies such as analyzing data, drawing contrasts, and making connections. It is not enough, however, to expand mathematics assessment to take in a broader spectrum of an individual student's competence. In real-world settings, knowledge is sometimes constructed in group settings rather than in individual exploration. Learning mathematics is frequently optimized in group settings, and assessment of that learning must reflect the value of group interaction.[13]

Some mathematics teachers are using group work in instruction to model problem solving in the real world. They are looking for ways to assess what goes on in groups, trying to find out not only what mathematics has been learned, but also how the students have been working together. A critical issue is how to use assessments of group work in the grades they give to individual students. A recent study of a teacher who was using groups in class but not assessing the work done in groups found that her students apparently did not see such work as important.[14] Asked in interviews about mathematics courses in which they had done group work, the students did not mention this teacher's course. Group work, if it is to become an integral and valued part of mathematics instruction, must be assessed in some fashion. A challenge to developers is to construct some high-quality assessment tasks that can be conducted in groups and subsequently scored fairly.

Part of the construction of knowledge depends on the availability of appropriate tools, whether in instruction or assessment. Recent experimental National Assessment of Educational Progress (NAEP) tasks in science use physical materials for a mini-experiment students are asked to perform by themselves. Rulers, calculators, computers, and various manipulatives are examples from mathematics of some instructional tools that should be a part of assessment. If students have been using graphing calculators to explore trigonometric functions, giving them tests on which calculators are banned greatly limits the questions they can be asked and

Since the use of

manipulatives is

a critical part of

today's

mathematical

instruction, such

tools must be

part of today's

assessment.

consequently yields an incomplete picture of their learning. Similarly, asking students to find a function that best fits a set of data by using a computer program can reveal aspects of what they know about functions that cannot be assessed by other means. Using physical materials and technology appropriately and effectively in instruction is a critical part of learning today's mathematics and, therefore, must be part of today's assessment.

Reflecting Development of Competence   As students progress through their schooling, it is obvious that the content of their assessments must change to reflect their growing mathematical sophistication. When students encounter new topics in mathematics, they often cannot see how the unfamiliar ideas are connected to anything they have seen before. They resort to primitive strategies of memorization, grasping at isolated and superficial aspects of the topic. As learning proceeds, they begin to see how the new ideas are connected to each other and to what they already know. They see regularities and uncover hidden relationships. Eventually, they learn to monitor their thinking and can choose different ways to tackle a problem or verify a solution.[15] This scenario is repeated throughout schooling as students encounter new mathematics. The example below contains a description of this growth in competence that is derived from research in cognition and that suggests the types of evidence that assessment should seek.[16]

A full portrayal of competence in mathematics demands much more than measuring how well students can perform automated skills although that is part of the picture. Assessment should also examine whether students have managed to connect the concepts they have learned, how well they can recognize underlying principles and patterns amid superficial differences, their sense of when to use processes and strategies, their grasp and command of their own understanding, and whether

---

Indicators of Competence

- *Coherent knowledge.* Beginners' knowledge is spotty and shallow, but as proficiency develops, it becomes structured and integrated into prior knowledge.

- *Principled problem solving.* Novices look at the surface features of a task; proficient learners see the structure of problems as they represent and solve them.

- *Usable knowledge.* Experts have knowledge that is connected to the conditions in which it can be applied effectively. They know not only what to do but when to do it.

- *Attention-free and efficient performance.* Experts are not simply faster than novices, they are able to coordinate their automated skills with thinking processes that demand their attention.

- *Self-regulatory skills.* As people develop competence, they also develop skills for monitoring and directing their performance.

---

they can bring these skills and abilities together to produce smooth, proficient performance.

Providing Feedback and Opportunities to Revise Work

An example of how assessment results can be used to support learning comes from the Netherlands.[17] Eleventh-grade students were given regular 45-minute tests containing both short-answer and essay questions. One test for a unit on matrices contained questions about harvesting Christmas trees of various sizes in a forest. The students completed a growth matrix to portray how the sizes changed each year and were asked how the forest could be managed most profitably, given the costs of planting and cutting and the prices at which the trees were to be sold. They also had to answer the questions when the number of sizes changed from three to five and to analyze a situation in which the forester wanted to recapture the original distribution of sizes each year.

After the students handed in their solutions, the teacher scored them, noting the major errors. Given this information, the students retook the test. They had several weeks to work on it at home and were free to answer the questions however they chose, separately or in essays that combined the answers to several questions. The second chance gave students the opportunity not simply to redo the questions on which they were unsuccessful in the first stage but, more importantly, to give greater attention to the essay questions they had little time to address. Such two-stage testing essentially formalizes what many teachers of writing do in their courses, giving students an opportunity to revise their work (often more than once) after the teacher or other students have read it and offered suggestions. The extensive experience that writing teachers have been accumulating in teaching and assessing writing through extended projects can be of considerable assistance to mathematics teachers seeking to do similar work.[18]

During the two-stage testing in the Netherlands, students reflected on their work, talked with others about it, and got information from the library. Many students who had not scored well under time pressure—including many of the females—did much better under the more open conditions. The teachers could grade the students on both the objective scores from the first stage and the subjective scores from the second. The students welcomed the opportunity to show what they knew. As one put it

> Usually when a timed written test is returned to us, we just look at our grade and see whether it fits the number of mistakes we made. In the two-stage test, we learn from doing the task. We have to study the first stage carefully in order to do well on the second one.[19]

In the Netherlands, such two-stage tasks are not currently part of the national examination given at the end of secondary school, but some teachers use them in their own assessments as part of the final grade each year. In the last year of secondary school, the teacher's assessment is merged with the score on the national examination to yield a grade for each student that is used for graduation, university admission, and job qualification.

LEARNING FROM THE SCORING OF ASSESSMENTS

Assessment tasks that call for complex responses require scoring rubrics. Such rubrics describe what is most important about a response, what distinguishes a stronger response from a weaker one, and often what characteristics distinguish a beginning learner from one with more advanced understanding and performance. Such information, when shared between teacher and student, has critically important implications for the learning process.

Teachers can appropriately communicate the features of scoring rubrics to students as part of the learning process to illustrate the types of performance students are striving for. Students often express mystification about what they did inadequately or what type of change would make their work stronger. Teachers can use rubrics and sample work marked according to the rubric to communicate the goals of improved mathematical explication. When applied to actual student work, rubrics illustrate the next level of learning toward which a student may move. For example, a teacher may use a scoring rubric on a student's work and then give the student an opportunity to improve the work. In such a case, the student may use the rubric directly as a guide in the improvement process.

The example on the following page illustrates how a scoring rubric can be incorporated into the student material in an assess-

Teachers can use scoring guides to communicate the goals of improved mathematical performance.

ment.[20] The benefits to instruction and learning could be twofold.
The student not only can develop a clearer sense of quality mathematics on the task at hand but can develop more facility at self-
assessment. It is hoped that students can, over time, develop an
inner sense of excellent performance so that they can correct their
own work before submitting it to the teacher.

Incorporating a Scoring Rubric

Directions for Students

Today you will take part in a mathematics problem-solving assessment. This means that you will be given
one problem to solve. You will have thirty (30) minutes to work on this problem. **Please show all your
work.** Your paper will be read and scored by another person — someone other than your teacher. Please
be sure to make it clear to the reader of your paper how you solved the problem and what you were
thinking. The person who will read you paper will be looking mainly for these things:

1. How well you **understand the problem** and the kind of math you use.
2. How well you can correctly use **mathematics**.
3. How well you can use **problem-solving strategies** and **good reasoning**.
4. How well you can **communicate** your mathematical ideas and your solution.

Your paper will receive a score for each of these. You will do all your work here in class on the paper
provided and you may use manipulatives or a calculator to work on your problem.

Guide to Completing the Problem

1. Conceptual understanding of the problem
   - ☐ I used diagrams, pictures, and symbols to explain my work.
   - ☐ I used all the important information to correctly solve the problem.
   - ☐ I have thought about the problem carefully and feel as if I know what I'm talking about.

2. Procedural knowledge
   - ☐ I used appropriate mathematical computations, terms, and formulas.
   - ☐ I correctly solved and checked my solution to the problem.
   - ☐ I used mathematical ideas and language precisely.
   - ☐ I checked my answer for correctness.

3. Problem-solving skills & strategies
   - ☐ I looked for other possible ways to solve the problem.
   - ☐ I used problem solving skills/strategies that showed some good reasoning.
   - ☐ My work is clear and organized.

4. Communication
   - ☐ I communicated clearly and effectively to the reader.
   - ☐ In my solution, one step seems to flow to the next.
   - ☐ I clearly used mathematics vocabulary and terminology.
   - ☐ My sentences make sense and there are no words left out.

The rubrics can be used to inform the student about the scoring criteria before he or she works on a task. The rubric can also be used to structure a classroom discussion, possibly even asking the students to grade some (perhaps fictional) answers to the questions. In this way, the students can see some examples of how responses are evaluated. Such discussions would be a purely instructional use of an assessment device before the formal administration of the assessment.

STIMULATING MOTIVATION, INTEREST, AND ATTENTION

Because assessment has the potential to affect the learning process substantially, it is important that students do their best when being assessed. Students' motivation to perform well on assessments has usually been tied to the stakes involved. Knowing that an assessment has a direct bearing on a semester grade or on placement in the next class—that is, high personal stakes—has encouraged many students to display their best work. Conversely, assessments to judge the effectiveness of an educational program where results are often not reported on an individual basis carry low stakes for the student and may not inspire students to excel. These extrinsic sources of motivation, although real, are not always consonant with the principle that assessment should support good instructional practice and enhance mathematics learning. Intrinsic sources of motivation, such as interest in the task, offer a more fruitful approach.

Students develop an interest in mathematical tasks that they understand, see as relevant to their own concerns, and can manage. Recent studies of students' emotional responses to mathematics suggest that both their positive and their negative responses diminish as tasks become familiar and increase when the tasks are novel.[21] Because facility at problem solving includes facility with unfamiliar tasks, the regular use of nonroutine problems must become a part of instruction and assessment.

In some school districts, educational leaders are experimenting with nonroutine assessment tasks that have instructional value in themselves and that seem to have considerable interest for the students. Such a problem was successfully tried out with fifth-grade students in the San Diego City School District in 1990 and has

Intrinsic sources of motivation offer a fruitful approach to encourage students to perform well.

subsequently been used by other districts across the country to assess instruction in the fifth, sixth, and seventh grades. The task is to help the owner of an orange grove decide how many trees to plant on each acre of new land to maximize the harvest.[22] The yield of each tree and the number of trees per acre in the existing grove are explained and illustrated. An agronomist consultant explains that increasing the number of trees per acre decreases the yield of each tree and gives data the students can use. The students construct a chart and see that the total yield per acre forms a quadratic pattern. They investigate the properties of the function and answer a variety of questions, including questions about extreme cases.

The assessment can serve to introduce a unit on quadratic functions in which the students explore other task situations. For example, one group of sixth-grade students interviewed an elementary school principal who said that when cafeteria lunch prices went up, fewer students bought their lunches in the cafeteria. The students used a quadratic function to model the data, orally reported to their classmates, and wrote a report for their portfolios.

Sixth-grade students can be successful in investigating and solving interesting, relevant problems that lead to quadratic and other types of functions. They need only be given the opportunity. Do they enjoy and learn from these kinds of assessment activities and their instructional extensions? Below are some of their comments.

Students' Comments on the Orange Tree Task

I Liked this task very much. I thought It was Interesting To see how the prices got higher and Lower. I had never heard of "Point of diminishing returns" but now I understand what that means. I wish we had done more this year.

I thought that this was very very fun I wish we could of done lots of these during the year. I thought this was the most fun kind of math so I have absolutly no complane<u>ts</u>

I thing this problem was kindof fun. And I liked it because it made me think a little harder than most problems we do. And it was interesting the way have to look up words in the dictionary

---

It is worth noting that the level of creativity allowable in a response is not necessarily tied to the student's level of enjoyment of the task. In particular, students do not necessarily value assessment tasks in which they have to produce responses over tasks in which they have to choose among alternatives. A survey in Israel of junior high students' attitudes toward different types of tests showed that although they thought essay tests reflected their knowledge of subject matter better than multiple-choice tests did, they preferred the multiple-choice tests.[23] The multiple-choice tests were perceived as being easier and simpler; the students felt more comfortable taking them.

ASSESSMENT IN SUPPORT OF INSTRUCTION

If mathematics assessment is to help students develop their powers of reasoning, problem solving, communicating, and connecting mathematics to situations in which it can be used, both mathematics assessment and mathematics instruction will need to change in tandem. Mathematics instruction will need to better use assessment activities than is common today.

Too often a sharp line is drawn between assessment and instruction. Teachers teach, then instruction stops and assessment occurs. Results of the assessment may not be available in a timely or useful way to students and teachers. The learning principle implies that "even when certain tasks are used as part of a formal, external assessment, there should be some kind of instructional follow-up. As a routine part of classroom discourse, interesting problems should be revisited, extended, and generalized, whatever their original sources." [24]

When the line between assessment and instruction is blurred, students can engage in mathematical tasks that not only are meaningful and contribute to learning, but also yield information the student, the teacher, and perhaps others can use. In fact, an oft-stated goal of reform efforts in mathematics education is that visitors to classrooms will be unable to distinguish instructional activities from assessment activities.

INTEGRATING INSTRUCTION AND ASSESSMENT

The new Pacesetter™ mathematics project illustrates how instruction and assessment can be fully integrated by design.[25] Pacesetter is an advanced high school mathematics course being developed by the College Board. The course, which emphasizes mathematical modeling and is meant as a capstone to the mathematics studied in high school, integrates assessment activities with instruction. Teachers help the students undertake case studies of applications of mathematics to problems in fields, such as industrial design, inventions, economics, and demographics. In one activity, for example, students are provided with data on the population of several countries at different times and asked to develop mathematical models to make various predictions. Students answer questions about the models they have devised and tackle more extended tasks that are written up for a portfolio. The activity allows the students to apply their knowledge of linear, quadratic, and exponential functions to real data. Notes for the teacher's guidance help direct attention to opportunities for discussion and the interpretations of the data that students might make under various assumptions.

Portfolios are sometimes used as the method of assessment; a sample of a student's mathematical work is gathered to be graded by the teacher or an outside evaluator.

> This form of assessment involves assembling a portfolio containing samples of students' work that have been chosen by the students themselves, perhaps with the help of their teacher, on the basis of certain focused criteria. Among other things, a mathematics portfolio might contain samples of analyses of mathematical problems or investigations, responses to open-ended problems, examples of work chosen to reflect growth in knowledge over time, or self-reports of problem-solving processes learned and employed. In addition to providing good records of individual student work, portfolios might also be useful in providing formative evaluation information for program development. Before they can be used as components of large-scale assessment efforts, however, consistent methods for evaluating portfolios will need to be developed.[26]

Of course the quality of student work in a portfolio depends largely on the quality of assignments that were given as well as on

An oft-stated goal of reform is that visitors to classrooms will be unable to distinguish instructional activities from assessment activities.

the level of instruction. At a minimum, teachers play a pivotal role in helping students decide what to put into the portfolio and informing them about the evaluation criteria.

The state of Vermont, for example, has been devising a program in which the mathematics portfolios of fourth- and eighth-grade students are assessed;[27] other states and districts are experimenting with similar programs. Some problems have been reported in the portfolio assessment process in Vermont.[28] The program appears to hold sufficient merit, however, to justify efforts under way to determine how information from portfolios can be communicated outside the classroom in authoritative and credible ways.[29]

The trend worldwide is to use student work expeditiously on instructional activities directly as assessment. An example from England and Wales is below.[30]

Assessment can be integrated with classroom discourse and activity in a variety of other ways as well: through observation, questioning, written discourse, projects, open-ended problems, classroom tests, homework, and other assignments.[31] Teachers need to be alert to techniques they can use to assess their students' mathematical understanding in all settings.

The most effective ways to identify students' methods are to watch students solve problems, to listen to them explain how the problems were solved, or to read their written explanations. Students should regularly be asked to explain their solution to a problem. Each individual cannot be asked each day, but over time the teacher can get a reading on each student's understanding and proficiency. The teacher needs to keep some

---

Coursework Assessment

As part of a new course in England and Wales, students aged 16 to 19 years are assessed through an externally marked final examination, tests given at the end of each unit of approximately 1 month's duration, and work done during the course. Each unit of coursework consists of a practical investigation extending throughout the unit and two short investigations of about 2 hours each. At the end of the course, 20 percent of each student's grade is based on the coursework and 80 percent is based on unit test and final examination scores. The coursework investigations are chosen from a bank provided by the examination board. Certain investigations are discussed in the text materials and are not used for assessment. Students usually work in groups during an investigation, but then each student writes an individual report to be marked by the teacher according to a set of criteria previously explained to the students.

For example, students in one class worked on the problem of finding a model for the motion of a ball rolling along an inclined plane. The data were collected and discussed in groups. Some students contributed greatly to the discussion; others did not. Although all those in the group had the benefit of the common work, the written reports clearly showed who had understood the problem and who had not.

From a Teacher-Constructed Seventh-Grade Japanese Semester Examination

Number game

  1. Choose a positive number and add 3.
  2. Multiply the result of (1) by 2.
  3. Subtract 3 from the result of (2).
  4. Multiply the result of (3) by 5.

If the result of (4) is 155, what is the original number? How did you find it? Explain how to find it.

Analysis

  The teacher analyzed students' explanations and found seven types of meaningful responses concerning the use of letters, as follows:

  • Uses a literal expression (roughly) and tries to explain by transformation
  • Explains by using numbers but does not depend on the numbers from the viewpoint of content
  • Explains by depending on numbers, but cannot detach from numbers
  • Finds a relation inductively
  • Explains by figures
  • Explains by language
  • Finds by the reverse process

The teacher evaluated each student according to these categories. Usually, it is difficult to carry out this type of analysis on a semester examination, since there is too little time.   But if it is carried out, the result is useful not only for assigning a grade but also for obtaining instructional feedback.

---

record of students' responses. Sunburst/Wings for Learning,[32] for example, recently produced the *Learner Profile*™, a hand-held optical scanner with a list of assessment codes that can be defined by the teacher. Useful in informal assessments, a teacher can scan comments about the progress of individual students while walking around the classroom.

  Elaborate schemes are not necessary, but some system is needed. A few carefully selected tasks can give a reasonably accurate picture of a student's ability to solve a range of tasks.[33]   An example of a task constructed for this purpose appears above.[34]

USING ASSESSMENT RESULTS FOR INSTRUCTION

  The most typical form of assessment results have for decades been based in rankings of performance, particularly in mandated assessment. Performances have been scored most typically by counting the number of questions answered correctly and comparing scores for one individual to that for another by virtue of their relative percentile rank. So-called norm referenced scores have concerned educators for many years. Although various criticisms on norm referencing have been advanced, the central educational concern is that such information is not sufficiently helpful to improve instruction and learning and may, in fact, have counterproductive educational implications. In the classroom setting, teachers and students need to know what students understand well, what they understand less well, and what the next learning steps need to be. The relative rankings of students tested may have uses outside the classroom context, but within that context, the need is for forms of results helpful to the teaching and learning process.

To plan their instruction, for example, teachers should know about each student's current understanding of what will be taught. Thus, assessment programs must inform teachers and students about what the students have learned, how they learn, and how they think about mathematics. For that information to be useful to teachers, it will have to include an analysis of specific strengths and weaknesses of the student's understanding and not just scores out of context.

To be effective in instruction, assessment results need to be timely.[35] Students' learning is not promoted by computer printouts sent to teachers once classes have ended for the year and the students have gone, nor by teachers who take an inordinate amount of time to grade assessments. In particular, new ways must be found to give teachers and students alike more immediate knowledge of the students' performance on assessments mandated by outside authorities so that those assessments—as well as the teacher's own assessments—can be used to improve learning. Even when the central purpose of an assessment is to determine the accomplishments of a school, state, or nation, the assessment should provide reports about their performance to the students and teachers involved. School time is precious. When students are not informed of their errors and misconceptions, let alone helped to correct them, the assessment may have both reinforced misunderstandings and wasted valuable instructional time.

When the form of assessment is unfamiliar, teachers have a particular responsibility to their students to tell them in advance

Assessment programs must inform teachers and students about what the students have learned, how they learn, and how they think about mathematics.

how their responses will be evaluated and what criteria will be used. Students need to see examples of work a priori that does or does not meet the criteria. Teachers should discuss sample responses with their students. When the California Assessment Program first tried out some open-ended questions with 12-grade students in its 1987-1988 Survey of Academic Skills, from half to three-fourths of the students offered either an inadequate response or none at all. The Mathematics Assessment Advisory Committee concluded that the students lacked experience expressing mathematical ideas in writing.[36] Rather than reject the assessment, they concluded that more discussion with students was needed before the administration of the assessment to describe what was expected of them. On the two-stage tests in the Netherlands, there were many fewer problems in scoring the essays when the students knew beforehand what the teacher expected from them.[37] The teacher and students had negotiated a kind of contract that allowed the students to concentrate on the mathematics in the assessment without being distracted by uncertainties about scoring.

· ·

ASSESSMENT IN SUPPORT OF TEACHERS

The new visions of mathematics education requires teachers to use strategies in which they function as learning coach and facilitator. Teachers will require support in several ways to adopt these new roles. First, they will need to become better diagnosticians. For this, they will need "... simple, valid procedures that enable [them] to access and use relevant information in making instructional decisions"; "assessment systems [that] take into account the conceptualizations of learning, teaching, and the curriculum that are held by teachers"; and systems that "enable teachers to share assessment data with students and to involve students in making instructional decisions." [38] Materials should be provided with the assessments developed by others that will enable teachers to use assessment tasks productively in their instruction. Help should be given to teachers on using assessment results to encourage students to reflect on their work and the teachers to reflect on theirs.

Teachers will require assistance in using assessments consonant with today's vision of mathematics instruction. The Classroom Assessment in Mathematics (CAM) Network, for example, is an electronic network of middle school teachers in seven urban centers

Teachers will require assistance in using assessments consonant with today's vision of mathematics instruction.

who are designing assessment tasks and sharing them with one another.[39] They are experimenting with a variety of new techniques and revising tasks to fit their teaching situation. They see that they face some common problems regarding making the new tasks accessible to their students. Collaborations among teachers, whether through networks or other means, can assist mathematics teachers who want to change their assessment practice. These collaborations can start locally or be developed through and sponsored by professional organizations. Publications are beginning to appear that can help teachers assess mathematics learning more thoroughly and productively.[40]

There are indications that using assessments in professional development can help teachers improve instruction. As one example, Gerald Kulm and his colleagues recently reported a study of the effects of improved assessment on classroom teaching:[41]

> We found that when teachers used alternative approaches to assessment, they also changed their teaching. Teachers increased their use of strategies that have been found by research to promote students' higher-order thinking. They did activities that enhanced meaning and understanding, developed student autonomy and independence, and helped students learn problem-solving strategies.[42]

This improvement in assessment, however, came through a substantial intervention: the teachers' enrollment in a three-credit graduate course. However, preliminary reports from a number of professional development projects such as CAM suggest that improved teaching practice may also result from more limited interventions.

Scoring rubrics can also be a powerful tool for professional development. In a small agricultural county in Florida, 30 teachers have been meeting on alternate weekends, attempting to improve their assessment practice.[43] The county has a large population of migrant workers, and the students are primarily of Mexican-American descent. The teachers, who teach mathematics at levels from second-grade arithmetic to calculus, are attempting to spend less time preparing the students to pass multiple-choice standardized tests. Working with a consultant, they have tried a variety of new tasks and procedures. They have developed a much greater respect for how assessments may not always tap learning. They found, for

Collaborations
with others
can assist
mathematics
teachers who
want to change
their assessment
practice.

example, that language was the biggest barrier. For students who were just learning English requests such as "discuss" or "explain" often yield little information. The teacher may need, instead, to ask a sequence of questions: "What did you do first?" "Why did you do that?" "What did you do next?" "Why?" and so on. Working with various tasks, along with the corresponding scoring rubrics, the teachers developed a new appreciation for the quality of their students' mathematical thinking.

Advanced Placement teachers have reported on the value of the training in assessment they get from the sessions conducted by the College Board for scoring Advanced Placement Tests.[44] These tests include open-ended responses that must be scored by judges. Teachers have found that the training for the scoring and the scoring itself are useful for their subsequent teaching of the courses because they focus attention on the most important features and lead to more direct instruction on crucial areas of performance that were perhaps ignored in the past.

Assessment tasks and rubrics can be devices that teachers use to communicate with parents and the larger community to obtain their support for changes in mathematics education. Abridged versions of the rubrics—accompanied by a range of student responses—might accomplish this purpose best. Particularly when fairly complex tasks have been used, the wider audience will benefit more from a few samples of actual student work than they will from detailed descriptions and analyses of anticipated student responses.

Teachers are also playing an active role in creating and using assessment results. In an increasing number of localities, assessments incorporate the teacher as a central component in evaluating results. Teachers are being recognized as rich sources of information about what students know and can do, especially when they have been helped to learn ways to evaluate student performance. Many students' anxiety about mathematics interferes with their test performance; a teacher can assess students informally and unobtrusively during regular instruction. Teachers know, in ways that test constructors in distant offices cannot, whether students have had an opportunity to learn the mathematics being assessed and whether they are taking an assessment seriously. A teacher can talk with students during or after an assessment, to find out how they inter-

Assessment tasks and rubrics can be devices for communicating with parents and the larger community.

preted the mathematics and what strategies they pursued. Developers of external assessment systems should explore ways of taking the information teachers can provide into account as part of the system.

Teachers are rich

sources of

information

about what

students know

and can do.

In summary, the learning principle aims to ensure that assessments are constructed and used to help students learn more and better mathematics. The consensus among mathematics educators is that assessments can fulfill this expectation to the extent that tasks provide students opportunities to extend their knowledge, are consonant with good instruction, and provide teachers with an additional tool that can help them to become better facilitators of student learning. These are new requirements for assessment. Some will argue that they are burdensome, particularly the requirement that assessments function as learning tasks. Recent experience—described below and elsewhere in this chapter—indicates this need not be so, even when an assessment must serve an accountability function.

The Pittsburgh schools, for example, recently piloted an auditing process through which portfolios developed for instructional uses provided "publicly acceptable accountability information." [45] Audit teams compsing teachers, university-based researchers, content experts, and representatives of the business community evaluated samples of portfolios and sent a letter to the Board of Education that certified, among other things, that the portfolio process was well defined and well implemented and that it aimed at success for all learners, challenged teachers to do a more effective job of supporting student learning, and increased overall system accountability.

There is reason to believe, therefore, that the learning principle can be honored to a satisfactory degree for both internal and external assessments.

ENDNOTES

[1] National Council of Teachers of Mathematics, *Curriculum and Evaluation Standards for School Mathematics* (Reston, VA: Author, 1989), 196.

[2] This statistic was compiled by using information from Edward D. Roeber, "Association of State Assessment Programs: Annual Survey of America's Large-Scale Assessment Programs" (Unpublished document, Fall 1991).

[3] Edward A. Silver and Patricia A. Kenney, "Sources of Assessment Information for Instructional Guidance in Mathematics" in Thomas A. Romberg, ed., *Reform in School Mathematics and Authentic Assessment*, in press; Edward A. Silver, Jeremy Kilpatrick, and S. Schlesinger, *Thinking Through Mathematics* (New York, NY: College Entrance Examination Board, 1990); Thomas A. Romberg, E. Anne Zarinnia, and Kevin F. Collis, "A New World View of Assessment in Mathematics," in Gerald Kulm, ed., *Assessing Higher Order Thinking in Mathematics* (Washington, D.C.: American Association for the Advancement of Science, 1990), 21-38; Thomas A. Romberg, "Evaluation: A Coat of Many Colors" (A paper presented at the Sixth International Congress on Mathematical Education, Budapest, Hungary, July 27-August 3, 1988), Division of Science, Technical and Environmental Education, UNESCO.

[4] Linda M. McNeil, "Contradictions of Control: Part 3, Contradictions of Reform," *Phi Delta Kappan* 69 (1998): 478-485.

[5] Lauren B. Resnick, National Research Council, Committee on Mathematics, Science, and Technology Education, *Education and Learning to Think* (Washington, D.C.: National Academy Press, 1987).

[6] Patricia Ann Kenney and Edward A. Silver, "Student Self-Assessment in Mathematics," in Norman L. Webb and Arthur Coxford, eds., *Assessment in the Mathematics Classroom*, 1993 NCTM Yearbook (Reston, VA: National Council of Teachers of Mathematics, 1993), 230.

[7] Thomas A. Romberg, "How One Comes to Know: Models and Theories of the Learning of Mathematics," in Mogens Niss, ed., *Investigations into Assessment in Mathematics Education: An ICMI Study* (Dordrecht, The Netherlands: Kluwer Academic Publishers, 1993), 109.

[8] Thomas A. Romberg and Thomas P. Carpenter, "Research on Teaching and Learning Mathematics: Two Disciplines of Scientific Inquiry," in Merlin C. Wittrock, ed., *Handbook of Research on Teaching*, 3d ed. (New York, NY: Macmillian, 1986), 851.

[9] *Education and Learning to Think*, 12.

[10] Nancy S. Cole, "Changing Assessment Practice in Mathematics Education: Reclaiming Assessment for Teaching and Learning" (Paper presented at the Conference on Partnerships for Systemic Change in Mathematics, Science, and Technology Education, Washington, D.C., 7 December 1992).

[11] This constructivist view of learning is becoming increasingly prevalent. Analyses of learning from a cognitive perspective point to the centrality of the learner's activity in acquiring understanding [see, for example, John R. Anderson, "Acquisition of Cognitive Skill, *Psychological Review*, 89 (1982): 396-406; and Y. Anzai and Herbert A. Simon, "The Theory of Learning by Doing" *Psychological Review* 86 (1979): 124-40). Classroom-based studies such as the ones cited earlier (Paul Cobb, Terry Wood, and Erna Yackel "Class-

rooms as Learning Environments for Teachers and Researchers," in Robert Davis, Carolyn Maher, and Nel Noddings, eds., *Constructivist Views on the Teaching and Learning of Mathematics*, monograph, no. 4 (Reston, VA: National Council of Teachers of Mathematics, 1990), 125-146; and Elizabeth Fennema, Thomas Carpenter, and Penelope Peterson "Learning Mathematics with Understanding: Cognitively Guided Instruction," in J. Brophy, ed., *Advances in Research in Teaching* (Greenwich, CT: JAI Press, 1989), 195-221]. Purely epistemological analyses [e.g., Ernst von Glasersfeld, "Learning as a Constructive Activity", in Claude Janvier, ed., *Problems of Representation in the Teaching and Learning of Mathematics* (Hillsdale, NJ: Lawrence Erlbaum Associates, 1987)], also lend credence to the conception of learners as constructors of their own knowledge.

[12] Lorrie A. Shepard, "Why We Need Better Assessments," *Educational Leadership*, 46:7 (1989), 7.

[13] There have been several reviews of the literature in this area, including Neil Davidson, "Small-Group Learning and Teaching in Mathematics: A Selective Review of the Literature, in R. Slavin et al., eds., *Learning to Cooperate, Cooperating to Learn* (New York, NY: Plenum, 1985), 211-230); Thomas L. Good, Catherine Mulryan, and Mary McCaslin "Grouping for Instruction in Mathematics: A Call for Programmatic Research on Small-Group Processes" in Douglas Grouws, ed., *Handbook of Research on Mathematics Teaching and Learning* (New York, NY: Macmillan, 1992); S. Sharan, "Cooperative Learning in Small Groups: Recent Methods and Effects on Achievement, Attitudes, and Ethinic Relations," *Review of Educational Research* 50 (1980), 241-271; R. Slavin, ed., *School and Classroom Organization* (Hillsdale, NJ: Lawrence Erlbaum Associates, 1989). *See also* Yvette Solomon, *The Practice of Mathematics* (London, England: Routledge, 1989), 179-187.

[14] Linda D. Wilson, "Assessment in a Secondary Mathematics Classroom" (Ph.D. diss., University of Wisconsin-Madison, 1993).

[15] Dedre Gentner and Albert L. Stevens, eds., *Mental Models* (Hillsdale, NJ: Lawrence Erlbaum Associates, 1981); Lauren Resnick and Wendy Ford, *The Psychology of Mathematics for Instruction* (Hillsdale, NJ: Lawrence Erlbaum Associates, 1981); Joseph C. Campione, Ann L. Brown, and Michael L. Connell, "Metacognition: On the Importance of Understanding What You Are Doing," in Randall I. Charles and Edward A. Silver, eds., *The Teaching and Assessing of Mathematical Problem Solving* (Reston, VA: Lawrence Erlbaum and National Council of Teachers of Mathematics, 1988), 93-114.

[16] Robert Glaser, "Cognitive and Environmental Perspectives on Assessing Achievement," in *Assessment in the Service of Learning: Proceedings of the 1987 ETS Invitational Conference* (Princeton, NJ: Educational Testing Service, 1988), 38-40.

[17] Jan de Lange, *Mathematics, Insight and Meaning: Teaching, Learning and Testing of Mathematics for the Life and Social Sciences* (Utrecht, The Netherlands: Rijksuniversiteit Utrecht, Vakgroep Onderzoek Wiskundeonderwijs en Onterwijscomputercentrum, 1987), 184-222.

[18] Vermont Department of Education, *Looking Beyond 'the Answer': The Report of Vermont's Mathematics Portfolio Assessment Program* (Montpelier, VA: Author, 1991); Jean Kerr Stenmark, *Assessment Alternatives in Mathematics: An Overview of Assessment Techniques that Promote Learning* (Berkeley, CA: University of California, EQUALS, 1989).

[19] *Mathematics, Insight and Meaning: Teaching, Learning and Testing of Mathematics for the Life and Social Sciences*, 207.

[20] Oregon Department of Education, *Student Directions: Guide to Completing the Problem* (Salem, OR: Author, 1991).

[21] Douglas B. McLeod, "Research on Affect in Mathematics Education: A Reconceptualization," in Douglas A. Grouws, ed., *Handbook of Research on Mathematics Teaching and Learning* (New York, NY: Macmillan, 1992), 578.

[22] Marilyn Rindfuss, ed., "Mr. Clay's Orange Orchard," Mathematics Performance Assessment, Form 1, *Integrated Assessment System Mathematics Performance Assessment Tasks* (San Antonio, TX: The Psychological Corporation, 1991).

[23] Moshe Zeidner, "Essay Versus Multiple-Choice Type Classroom Exams: The Student's Perspective," *Journal of Educational Research* 80:6 (1987), 352-358.

[24] National Research Council, Mathematical Sciences Education Board, *Measuring Up: Prototypes for Mathematics Assessment* (Washington, D.C.: National Academy Press, 1993), 11.

[25] The College Board, *Pacesetter: An Integrated Program of Standards, Teaching, and Assessment* (New York, NY: Author, 1992).

[26] Edward A. Silver, "Assessment and Mathematics Education Reform in the United States," *International Journal of Educational Research* 17:5 (1992), 497.

[27] *Looking Beyond 'The Answer'.*

[28] Daniel Koretz et al., *The Reliability of Scores from the 1992 Vermont Portfolio Assessment Program*, CSE Technical Report 355 (Los Angeles, CA: University of California, National Center for Research on Evaluation, Standards, and Student Testing, 1993).

[29] Pamela A. Moss et al., "Portfolios, Accountability, and an Interpretive Approach to Validity," *Educational Measurement: Issues and Practice* 11:3 (1992), 12-21.

[30] 31. Adapted from A. England, A. Kitchen, and J. S. Williams, *Mathematics in Action at Alton Towers* (Manchester, England: University of Manchester, Mechanics in Action Project, 1989).

[31] "Sources of Assessment Information for Instructional Guidance in Mathematics."

[32] Sunburst/Wings for Learning, *Learner Profile* (Pleasantville, New York: Author, 1993).

[33] In a sense this relates to the notion of generalizability, the extent to which inferences about performance on a totality of tasks can be inferred from performance on a subset. In the relatively informal milieu of internal assessment, of course, it is fairly easy for teachers to supplement an assessment with additional tasks if they are not convinced that they have sufficient data from which to make judgments. Nonetheless, the effectiveness of internal assessment is heavily dependent on the teacher's skill and acumen in task selection.

[34] Shinya Ohta, "Cognitive Development of a Letter Formula" (in Japanese), *Journal of Japan Society of Mathematical Education* 72 (1990):242-51, in Ezio Nagasaki and Jerry P. Becker, "Classroom Assessment in Japanese Mathematics Education" in Norman L. Webb and Arthur F. Coxford, eds., *Assessment in the Mathematics Classroom* (Reston, VA: National Council of Teachers of Mathematics, 1993), 40-53.

[35] R. L. Bangert-Drowns et al., "The Instructional Effect of Feedback in Test-Like Events," *Review of Educational Research*, 61:2 (1991), 213-238. This study reported a meta-analysis of 40 studies that showed that (a) immediate feedback is more effective than feedback that is delayed a day or more after a test, and (b) providing guidance about correct answers is more effective than feedback that merely informs students whether their answers were correct or not.

[36] California Assessment Program, *A Question of Thinking: A First Look at Students' Performance on Open-Ended Questions in Mathematics* (Sacramento, CA: California State Department of Education, 1989), 6.

[37] *Mathematics, Insight and Meaning*, 218.

[38] Margaret C. Wang, "The Wedding of Instruction and Assessment in the Classroom," in *Assessment in the Service of Learning: Proceedings of the 1987 ETS Invitational Conference* (Princeton, NJ: Educational Testing Service, 1988), 75.

[39] Maria Santos, Mark Driscoll, and Diane Briars, "The Classroom Assessment in Mathematics Network," in Norman L. Webb and Arthur Coxford, eds., *Assessment in the Mathematics Classroom*, 1993 NCTM Yearbook (Reston, VA: National Council of Teachers of Mathematics, 1993), 220-228.

[40] Examples include J.K. Stenmark, *Mathematics Assessment: Myths, Models, Good Questions, and Practical Suggestions* (Reston, VA: National Council of Teachers of Mathematics, 1991); *Assessment in the Mathematics Classroom*; *Measuring Up*; *Assessing Higher Order Thinking in Mathematics*; California Assessment Program, *A Sampler of Mathematics Assessment* (Sacramento, CA: California Department of Education, 1991); Judy Mumme, *Portfolio Assessment in Mathematics* (Santa Barbara, CA: California Mathematics Project, University of California, Santa Barbara, 1990).

[41] Gerald Kulm, "A Theory of Classroom Assessment and Teacher Practice in Mathematics (Symposium paper presented at the annual meeting of the American Educational Research Association, Atlanta, GA, 17 April 1993). Related papers at the same symposium were Bonita Gibson McMullen, "Quantitative Analysis of Effects in the Classroom"; Diane Scott, "A Teacher's Case of New Assessment"; James A. Telese, "Effects of Alternative Assessment from the Student's View."

[42] "A Theory of Classroom Assessment," 12.

[43] Gilbert Cuevas, personal communication, April 1993.

[44] The College Board, *An Invitation to Serve as a Faculty Consultant to the Advanced Placement Reading* (New York, NY: Author, 1993).

[45] Paul LeMahieu, "What We Know about Performance Assessments" Session (Presentation made at the annual conference of the National Center for Research on Evaluation, Standards, and Student Testing, Los Angeles, CA, 10 September 1992).

5 ASSESSING TO SUPPORT EQUITY AND OPPORTUNITY IN MATHEMATICS LEARNING

High-quality mathematics assessments must focus on equity concerns to ensure that all students are well served by assessment practices. This fundamental concept is embodied in the third educational principle

THE EQUITY PRINCIPLE
Assessment should support every student's opportunity to learn important mathematics.

The tendency in education has been to think about equity in terms of groups of children who by some definition are not well served by educational institutions. Quite often in these discussions the focus has been on groups that are defined in terms of status variables such as sex, ethnicity, native language, and socioeconomic indicators, because these factors influence access, expectations, and the systems of rewards. Increasingly, however, educators have recognized that although status characteristics of learners may be important, they are far less important than functional characteristics are in the design of instructional strategies.[1] Reforms in mathematics and other areas of schooling must aim to ensure that each student, regardless of achievement level or demographic characteristics, has the opportunity to study challenging subject matter. The example on the following page, a policy statement adopted by the National Council of Teachers of Mathematics (NCTM) in April 1990,[2] reflects this perspective on equity.

Some proponents of educational reform maintain that use of more flexible assessment methods and more performance-oriented tasks can produce fairer measures of *all* students' intellectual development.[3] Others have cautioned that new assessment content and methods alone can not assure equity:

Mathematics for *All* Students

In April 1990, the NCTM Board of Directors endorsed the following statement:

As a professional organization and as individuals within that organization, the Board of Directors sees the comprehensive mathematics education of every child as its most compelling goal.

By every child we mean specifically

- students who have been denied access in any way to educational opportunities as well as those who have not;
- students who are African American, Hispanic, American Indian, and other minorities as well as those who are considered to be a part of the majority;
- students who are female as well as those who are male; and
- students who have not been sucessful in school and in mathematics as well as those who have been successful.

> The call for performance assessment did not derive from concerns for fairness or equal access . . . Well designed assessment systems might act as a lens to more clearly reveal existing and ongoing inequalities, and to inform policy and practice . . . but only if explicitly designed to do so in all its parts.[4]

Designing for equity requires conscientious rethinking of not just what we assess and how we do it, but how different individuals and groups are affected by assessment design and procedures. It also requires conscientious attention to how we use assessment results to make decisions about individual children and the schools they attend.

DEVELOPING ASSESSMENTS TO INCREASE EQUITY

To meet the equity principle, tasks must be designed to give children a sense of accomplishment, to challenge the upper reaches of each child's mathematical understanding, and to provide a window on each student's mathematical thinking. Just as good instruction accommodates differences in the ways learners construct knowledge, good assessments accommodate differences in the ways that students think about mathematics and display mathematical understanding. Although all students are to be assessed on important mathematical concepts and skills, in accord with the content and learning principles, the equity principle implies that assessments must be sufficiently flexible to allow all students to show what they know and can do.

DESIGN FEATURES

Some ways of accommodating differences among learners include permitting multiple entry and exit points in an assessment and allowing students to respond in ways that reflect different levels of mathematical knowledge or sophistication. These design characteristics are critical to assessment equity just as they are critical to the content and learning principles.

Consider the two apparently parallel problems illustrated below and on the following page.[5] Close inspection indicates that the first problem may be less accessible to many students. In the first mosaic there is no indication of how the colored tiles are arranged to form the picture. Therefore, to find how many of the different colored tiles would be needed for the enlarged picture, one must already know the general proposition that doubling the length of a figure quadruples its area. On the other hand, the figure in the second problem, with its explicit specification of the color of each part, opens alternative avenues for approaching the task. For example, the student might draw a sketch

<table>
<tr><td>

Pythagoras—A Mosaic Problem

Your school's math club has designed a fancy mosaic in honor of Pythagoras, whose work united algebra and geometry. The mosaic includes a picture of the Pythagorean Theorem and its algebraic statement, as indicated below. The figure appears inside a square.

You have constructed a 2' x 2' scale model of the mosaic, which used 120 red mosaic pieces, 200 blue mosaic pieces, and 150 yellow pieces.

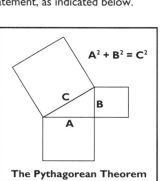

$A^2 + B^2 = C^2$

The Pythagorean Theorem

(a) How many red, blue, and yellow pieces would be required for a 4' x 4' mosaic? Explain.

(b) How many red, blue, and yellow pieces would be required for a 6' x 6' mosaic? Explain.

(c) The school's principal likes the model of the mosaic so much that she's willing to have you build a big version that will go on permanent display in the auditorium. The only thing—she hasn't yet decided how big the mosaic should be. She wants the math club to write her a letter explaining how many tiles of each color she'll need, as soon as she makes up her mind how big the side of the square should be. Write the letter, providing an explanation of how the number of tiles can be calculated, and also why your answer is correct.

</td></tr>
</table>

Just as good

instruction

accommodates

differences in

the ways

learners

construct

knowledge, good

assessments

accommodate

differences in

the ways that

students think

about and

display

mathematical

understanding.

Tri-Tex Logo—A Mosaic Problem

Tri-Tex Corporation has the following logo. It's going to put a large-scale mosaic of the logo on the side of its corporate headquarters building.

A 2' x 2' scale model of the logo uses: 144 yellow tile pieces
144 blue tile pieces
288 red tile pieces

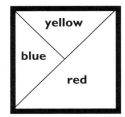

When the logo is placed on the side of the building it will also have a border that consists of a single band of long black tiles. It takes 40 of the black tiles to make a border for the 2' x 2' scale model.

- How many tiles of each color will the purchasing department have to order if the full-scale mosaic is 4' x 4'? What if it's 6' x 6'?

- Tri-Tex's president says he'll decide later just how big the mosaic will be, but he wants everything set up in advance so the purchasing department can send out the order as soon as he makes up his mind. Your job is to write a two-part memo to the purchasing department.

(a) In part 1 of the memo, tell how the purchasing department can do a simple calculation to find out how many yellow, blue, red, and black tiles they need to purchase, for any size design. The instructions should be as simple and direct as possible.

(b) In part 2 of the memo, explain how you arrived at the answer you did, so the people in the purchasing department can understand why your instructions give the right answer.

of the enlarged figure and discover how many congruent copies of the original figure will be needed to cover it. In fact, by using an informal, experimental approach to the task, the student has the opportunity to gain insight into the relation between length and area, thus potentially learning some important mathematics in the process of doing the task.

As the preceding illustration suggests, assessments designed with multiple entry and exit points permit each student—even those who had mastered only the most elementary knowledge and skills tapped by a problem—to experience some success. However, this goal is not always easily accomplished. Consider the response on the following page of a Ute Indian student to an assessment task.[6] Rather than focusing on the mathematical question asked, the student responded to the hypothetical nature of its premise.

Children who live in harsh environments in the inner cities and in remote rural areas often respond to hypothetical problems in much the same way. Hypothetical situations representing options that do not exist in their daily lives may be treated as meaningless or nonsensical. Therefore, their responses may reveal little about their

A Cultural Convention

An elementary school student from the Northern Ute Indian reservation in Utah was interviewed as he solved some word problems. He was asked to determine how much his brother would have to spend on gasoline if he wanted to drive his truck from the reservation to Salt Lake City. Instead of estimating (or generalizing) a response, or attempting to calculate an answer based on the information presented in the request, the student responded quite simply: "My brother does not have a pickup."

He was assessing the truth value of the conditions presented by the problem. Such assessments are a common part of everyday life of this reservation. Decisions regarding personal conduct as well as innovations and changes affecting tribal membership as a whole are made only if the proposed line of action seems consistent with recognized cultural convention. Proposals that conflict with the way things are may be briefly considered, then set aside for action at some later time, or may simply be given no serious discussion at all.

understanding of the mathematical problem posed or the mathematical knowledge and skills required by the problem.

As assessments become more complex and more connected to real-world tasks, there is a greater chance that the underlying assumptions and views may not apply equally to all students, particularly when differences in background and instructional histories are involved. In such instances special care will be needed to determine whether differences in performance stem from differences in unrelated characteristics or genuinely reflect differences in mathematical development.

SUPPORTING STRATEGIES

When assessments are administered by the teacher in a classroom setting to gauge student progress or diagnose a particular difficulty, differences in students' backgrounds that might affect results are likely to be known. In such cases skillful probing can be used to clarify the basis of the students' responses. Questions can be reframed in ways that better reflect students' understanding of the mathematics required.

In many assessment settings, particularly those used for accountability, it is less possible to accommodate differences in students' ways of understanding and responding to a task. To help overcome this difficulty, it is important that varied perspectives be represented at every phase of assessment development and use. The teams that create the intellectual frameworks from which specific tasks are generated should include experts with different views and life experiences. Broad-based representation is also vital on the research teams that pose study

The evolving agenda for research on assessment must reflect as much concern for equity as for content, learning, and technical issues.

questions, devise data collection protocols and strategies, and oversee analyses and reporting of results. It is equally important in pilot research: sites should be selected to provide realistic variation among students, teachers, and schools. The raters who evaluate student responses should also represent varied backgrounds.[7]

The evolving agenda for research on assessment must reflect explicit concern for equity as well. Several projects of major national significance have taken steps in this direction. The New Standards Project (NSP), for example, has created special equity initiatives to ensure that the standards and tasks adopted reflect the experiences and aspirations of varied groups. NSP has conducted focus groups with local and national advocacy representatives to gain broad perspectives on how equity concerns might be addressed. It has also established advisory groups of scholars and practitioners with expertise in culture, gender, and language issues to advise the project on development of assessment tasks and procedures. In addition, a special advisory group was created to address issues involved in administering appropriate assessment tasks and procedures for students who are not native speakers of English.[8]

The Center for Research, Evaluation, and Student Testing (CRESST) at the University of California at Los Angeles also has special equity initiatives. CRESST created a task force of consultants and scholars to investigate the equity challenges confronting the assessment community. The task force has a broad agenda, encompassing very practical issues as well as a range of theoretical issues and basic research problems:

- What dimensions of diversity have relevance for education and the assessment of educational achievement?

- How do group identity and status influence learning of varied cultural groups and their responses to assessment probes?

- How can problems of reliability, validity, and calibration posed by cultural diversity be addressed?

- How can procedures and criteria be developed so that appropriate and inappropriate responses are defined that are sensitive to differences among the groups being assessed?

- On what kinds of tasks do specific groups tend to excel or have difficulty?[9]

INTERPRETING ASSESSMENT RESULTS

Despite efforts such as those described above and the hopes of the educational reform community that new assessments will be fairer to every student than traditional forms of testing were, preliminary research does not confirm the corollary expectation that group differences in achievement will disappear. Surely, there are many anecdotal reports from classroom teachers and researchers of some students who perform quite poorly on traditional tests exhibiting advanced reasoning and mathematical understanding on more flexible assessments. However, several studies suggest that differences in average group performance are essentially unchanged with the switch from multiple-choice to open-ended assessments. Still other research suggests that group differences might be magnified when performance assessment tasks are used.[10] Although the results may vary depending on the content area assessed, the nature of the specific tasks used, and whether group or individual performance is the focus of comparisons, it seems clear that reformers' hopes must be balanced by a spirit of empiricism: there is much more to be learned about how changes in assessment will affect long-standing group differences, one facet of equity concerns.

In any event, equity implies a commitment to seek and explore the sources of any systematic performance differences that may be observed on new assessments. It is important to know whether observed differences among groups reflect genuine differences in mathematical development or can be attributed to factors such as those identified in earlier chapters as special challenges for assessment design:

- The *context* in which a task is presented can influence mathematics performance.[11] The words and concepts used, the expressions of knowledge required or permitted, will always reflect some particular set of values or life experiences. Not all children will be equally familiar with any given context. Use of genuinely novel contexts or pre-assessment experience with a given context will allow students to enter on an equitable footing.

- *Language* can be a potent factor in mathematics assessments.[12] Better readers may more easily grasp what they

Equity implies a commitment to seek and explore the sources of systematic performance differences that may be observed on new assessments.

are being asked to do on an assessment task. Better writers may more often present their ideas well, even when of marginal mathematical quality. Because communication is a legitimate facet of mathematical power, some balance must be sought in determining how to weigh students' general command of language in evaluating their expressions of mathematical ideas.

- *Open-ended assessment procedures* may exacerbate the influence of resource inequities that exist in the real world. If parents or teachers participate in rather than support students' work, assessments may overestimate the student's mathematical fluency.

- *Rater influences* may distort judgments of performance on more open-ended tasks. Many factors can come into play: the rater's expertise and knowledge of the specific mathematics content of the task, the rater's interpretation of task requirements and scoring criteria, or even similarities or differences in the backgrounds of the rater and the student being evaluated.[13]

When any of these factors contribute to systematic differences among definable groups, assessment results have meanings or implications that are different for some groups than for others. Inferences based on results from such an assessment are said to be *biased*, that is, not equally valid for all groups.[14]

Solid documentation on the students assessed, the schools they attend, and the instruction they have received makes it possible to identify inequities in opportunity.

Solid documentation on the students assessed, the schools they attend, and the instruction they have received makes it possible to identify inequities in opportunity. Pilot testing provides an opportunity to ferret out biases and to try out potential remedies. Thorough documentation also is important when assessments are fully implemented so that appropriate cautions and conditions can be applied to interpretations and uses of the results.[15] Remedies for group differences depend on the source of the difference. Not all differences will be due to bias. Some, in fact, may be due to systematic differences in the quality of mathematics instruction provided. The following example outlines some of the kinds of information needed to form inferences about individuals and schools.[16]

Equity of Opportunity Indices

CRESST has identified the kinds of process information that will be needed about the quality of students' educational experiences to ensure that clear inferences can be drawn from assessment results:

- What evidence is there that the students have had opportunity to learn the assessed material? What evidence is there of the quality of those experiences? What is the cumulative experience of transient or frequently absent students?

- What evidence is there that poorly performing groups of students have been taught by teachers of the same quality, training, and experience as more successful students?

- What are the net educational resources available to students, including compensatory family and community benefits? Are comparable books, materials, and the other educational supports available across groups?

- Is there evidence that the affective environments of education are comparable? How safe are the schools? What size are the schools? How are individual rights and needs accommodated?

- How balanced is the exposure of students across the full range of desired educational outcomes, both measured and unmeasured?

● ●

USING ASSESSMENTS TO COMMUNICATE NEW EXPECTATIONS

Assessments represent an unparalleled tool for communicating the goals and substance of mathematics education reform to various stakeholders. Teachers, students, parents, policymakers, and the general public will need to understand clearly where mathematics reform will take America's children and why they should support the effort. Assessments can be enormously helpful in this reeducation campaign.

PROMOTING PUBLIC INVOLVEMENT

Increasingly, public dialogue is seen as an important strategy for ensuring that assessments are designed and used equitably.[17] The Urban District Assessment Coalition (UDAC), a joint effort of Boston College and the American Federation of Teachers, organized a series of conversations about assessment reform among various groups before designing a new assessment system for the Boston schools. Over several months, parents, teachers, administrators, and members of the business community worked to arrive at a

Public dialogue is an important strategy for ensuring that assessments are designed and used equitably.

consensus on the values, goals, and information needs the system would serve. Representatives of each group have been involved in all stages of the development process, from planning to scoring responses during the first round of pilot tests. That process is being replicated in other UDAC sites.[18] Another strategy for increasing public engagement involves making assessment blueprints public. CRESST has proposed that the public be given quite extensive information in blueprints,[19] as indicated in the example below.

Assessment Blueprints

CRESST views public reporting of assessment blueprints as an equity remedy for assessment design. They have recommended that the following questions be addressed in such blueprints:

- *World knowledge:* What common experiences and understandings are required of students to make sense of the task and to productively undertake its solution? Which of these are more or less likely for students of different cultural and socioeconomic backgrounds?

- *Prior knowledge:* What specific types of information or resources are essential for successful performance?

- *Language demands:* If the focus of the task is not language facility, are alternative options for display of understanding available for students with limited English proficiency?

- *Task structure and topics:* How were task structure and topics created? Is there reason to believe that all groups of children will be motivated by the topics provided? Are sufficient numbers of topics represented to draw safe inferences about the task domain of interest?

- *Scoring criteria:* What criteria will be used to judge student performance? Are these criteria specific enough to overcome biases potential in global ratings of performance?

- *Judges:* How are raters selected? Do they possess high degrees of relevant knowledge of the domains to be assessed? How have they been trained? Have raters' preferences for performances of ratees of like ethnicity been estimated?

- *Access:* Who is excluded from assessments and on what basis? How comparable are exclusion rules? What special provisions for access are available to students with special needs?

- *Equity of educational assessment settings:* What is the quality of educational experiences to which students have been exposed?

PROMOTING COMMUNITY UNDERSTANDING OF ASSESSMENT REFORM

For mathematics education reform to reach each child, parents and other community members will need to understand the new vision of school mathematics. Their involvement is needed to support the efforts of the teachers, school administrators, and others who, with them, share responsibility for the education of their children. Because many students are "doing well" or "showing improvement" in mathematics achievement as measured by many of the tests in use today, members of each school community will need to be assured that new assessments and the standards they reinforce are not arbitrary. They need to understand that "doing better" by old standards is no longer good enough to secure children's futures. Assessment tasks and their corresponding scoring criteria can be used to make this point quite convincingly. They illustrate that the knowledge and proficiency that students are being asked to demon- strate are of value to *all* students.

Supervisors in the San Francisco Unified School District have reported that working with assessments in PTA meetings helped parents in one community see exactly how the mathematics that their children will need differs from the mathematics the parents learned.[21] These sessions helped assure parents that much of the mathematics that they value (e.g., numbers and computation) remains important but that it is being learned and assessed as a part of meaningful problem solving and reasoning. After preliminary discussion of the kinds of mathematics students would need in the future and an overview of K–12 program content, parents met in small groups to discuss the mathematics required by several prob- lems. One of the more challenging problems is shown below.[22] Scoring rubrics and sample responses were also made available to the small groups.

The program coordinator reported several interesting reactions from parents. Many were awed by the kinds of problems that students are now expected to do and the sophisticated math- ematical problem solving and reasoning required. Most believed it important for their children to develop the kinds of skills the problem tapped. Most also believed their children could do the kind of work required but stressed the importance of students getting an early start in primary grades. Many parents asked questions about the quality of mathematics instruction at the school and sought reassurance that their children were getting the right kind of teaching. Most significantly, perhaps, no parent asked why children were not drilling on multiplication facts, a rare

> Parents and community members need to understand that doing well by old standards is no longer good enough to secure children's futures.

Multiplication Task Used in Parent Workshop

The five digits—1, 2, 3, 4, and 5—are placed in the boxes above to form a multiplication problem. If they are placed to give a maximum product, the product will fall between:

A. 10,000 and 22,000   B. 22,001 and 22,300
C. 22,301 and 22,400   D. 22,401 and 22,500

outcome at meetings that focus on curriculum content, according to the mathematics and science supervisor.

CHALLENGING TEACHERS' EXPECTATIONS

Working with

assessments

challenges

teachers'

expectations

about who can

learn important

mathematics.

Working with assessments and scoring criteria helps teachers crystallize their understanding of what content and teaching standards in mathematics and new performance expectations imply about what should be taught and how. Working with assessments also challenges teachers' expectations about who can learn important mathematics. Teachers sometimes discover that students who rarely make a good showing when tested in traditional ways truly shine on tasks that are more flexible and more engaging.

For example, teachers working with researchers at the Educational Testing Service have been surprised at which students do well and which do poorly on complex tasks. One of the tasks required students to explain how to set up an activity similar to a game of darts, although it involved tossing compact discs rather than darts.[23] As indicated in the eaxmple below, students were to write a description that included information on how the game would be set up, chances of winning, and expected profits.

The Compact Disc Tossing Game

For a popular carnival game, a player tosses a coin onto a game board that looks like a checker board. If the coin touches a line on the game board the player loses. If not, the player wins. Players get three throws for a dollar!

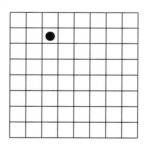

You've been asked to design a similar game for a fund-raising carnival at your school. For prizes, a local record store will sell up to 100 compact discs (CDs) to your class for $5 per disc. You can choose any CDs you want as long as the regular price of the CD is less than $20. To make the game more fun, you've decided to let players throw old scratched CDs rather than coins. Two sizes of CDs can be used (3-inch discs and 5-inch discs). So, the cost of three throws can depend on which size a player chooses to throw.

All plans for games must be approved by the carnival planning committee. You want to make as large a profit as possible, but if too few people win, people won't want to play the game. Write a plan to submit to the carnival planning committee that includes details about the size of the game board, the cost of throws, the chances of winning, and an estimate of the expected profits.

When low-performing middle school students were given an hour or more to work on the problem using physical models, several came up with solutions that were similar to, though less formal than, answers given by college-level mathematicians who spent 5 to 10 minutes on the task. The example below traces the conversation of three students' efforts to think through the problem. The boys began by drawing a game board on a large sheet of paper and cutting out several cardboard disks that were used to act out the game. They worked in this way for about 15 minutes before noticing that the size of the disk made a difference. Their conversation from that point follows.[24]

Dialogue of Three Students Working on the CD Tossing Game

Jamal: Mine [small] is better. Yours never works [too large].

[Five minutes passes as the boys tried several more disks, gradually making them smaller and smaller (so they could win more often).]

Bart: This would be easy if this thing [disk] was a dart.
Carl: Yah, but it's not a dart.
Bart: But what if it was. Look! Look at this. [Bart draws lots of little squares inside the original game board squares as shown in the figure below.] I win if my dart lands in here [referring to his disk as a dart, and pointing to the shaded squares in the diagram shown below.] And, I lose if the dart lands out here [pointing to the unshaded squares]. I've got this many chances to win, and this many to lose.
Jamal: Let's count 'em. [He counts, but not very carefully.] There are 300 here [pointing to the unshaded squares.] And 100 here [pointing to the shaded squares].
Bart: So I've got a 100 to 300 chance to win.
Carl: Yah, but what if you was Michael Jordan? He'd throw it like this [down the center of one row of squares]. He wouldn't throw it like this [from corner to corner].
Jamal: That's right! And, Tina, my sister, she's so dumb, she'd miss the whole thing.

Both the boys' solution and the mathematicians' solution were modeled in terms of the position of the discs' center relative to the "win" area of the game board. However, the university professors produced solutions based on the (generally false) assumption that the tosses were random but still good enough to hit the gameboard. In several important respects, however, the three students come up with a more elegant solution, as their letter to the carnival committee makes clear. Reproduced verbatim, it says:

Fairness

requires that

students

understand what

they are

expected to do

and what criteria

will be used to

judge their

performance.

This is a good game. Don't make the squares on the board too big. Make it about this size (a drawing is given). If your like Tina Jackson, your chances to lose are about 100 to 1. If your like Michael Jordan, your chances are about 50/50. Most people will be about 1 to 8. 500 people will come. About half will be like Tina. They won't play much. Only a few guys will be like Michael Jordan. They like to play a lot. So, make them quit after they win once. There are about 20 of these guys. They will all win some-times. Most people who play aren't like Tina or Jordan. About 200 will play. You should charge 1 dollar to play the game.[25]

The boys' solution reflected understanding of conditional probabilities and identified the factors that would determine success and profits. The boys who produced this solution also demon-strated exceptional mathematics capability when assessed on other problems that rewarded their use of real-life knowledge and skills. Although not all students with histories of low academic perfor-mance produced elegant solutions in response to challenging tasks, it was not unusual to discover at least two or three such students in a typical-size classroom. It also was not unusual to find cases in which high achievers, based on traditional measures, exhibited poorer performance on such problems.[26]

HELPING STUDENTS UNDERSTAND AND MEET NEW EXPECTATIONS

The equity principle requires that students be prepared for more challenging assessments. Fairness requires that students under-stand what they are expected to do and what criteria will be used to judge their performance. As noted earlier assessment tasks and scoring criteria provide good targets to teach to and good models of the kinds of mathematical thinking expected of all students, especially to those who, in the past, have been held to a lower standard.

USING ASSESSMENT RESULTS TO SUPPORT OPPORTUNITY

Although it is important to consider how differences in opportunity can affect assessment results, it is equally important to be aware of how assessment results can affect students' opportuni-ties to learn important mathematics. The equity principle implies that uses of assessment results be scrutinized for their impact on both students' opportunities to learn and schools' capacities to provide instruction consonant with the vision of mathematics education.

RETHINKING ASSESSMENT USE FOR SORTING AND GROUPING

Assessment has always been looked to as much for how it could improve educational opportunity as for how it might be used to sort and select. For every Galton who looked for quantitative differences between groups, there has been a Binet who wanted to help children in need of special education and who resisted attempts to reify intelligence and rank children accordingly. When inadequate attention has been paid to the policy context into which testing programs are introduced, however, many negative, unintended consequences have followed. Binet could never have anticipated that his mental tests would be used to support an American hereditarian theory of intelligence and its tragic aftermath.[27]

More recently, states and school districts introduced minimum competency tests to ensure that each student who was graduated or promoted learned essential subjects. Some of the anticipated benefits of these policies are being realized. One clear benefit was that differences in the achievement of African American and white students began to decline,[28] but there were unanticipated negative effects as well. When these programs were implemented along with sanctions for schools, many had punitive effects on low-achieving students and schools serving large numbers of those students. For example, in some cases, low-achieving students were retained in grade or identified for special education services and thus exempted from assessment. Low-performing schools had difficulty attracting and keeping good teachers with the threat of sanctions in force.[29] Another negative outcome was minimum standards and expectations, which resulted in schools emphasizing basic skills at the expense of the kinds of higher-order skills that students need to function in a technologically advanced society.[30]

A growing consensus among policymakers, assessment developers, and educators is that if assessments continue to be used to sort learners rather than to promote learning and to punish schools and teachers rather than to support them, new assessments will have the same effects as the old ones.[31] For assessments to help children rather than harm them and to promote better instruction

For assessments
to help children
and to promote
better
instruction
rational policies
must be
developed for
using
assessment
results.

rather than to manipulate test scores, rational policies must be developed for using assessments. Policies that promote equity may challenge well-entrenched school practice and the educational philosophies upon which it rests. Clearly, teachers' evaluations play a strong role in grouping and tracking decisions, but the sorting would not take place without shared beliefs to sustain it. The practice of sorting rests on a general societal preoccupation with individual differences in ability, particularly in ability to learn and achieve in mathematics. Assessment also plays a role in perpetuating the practice. The scientific mystique that surrounds tests and testing procedures helps to convince teachers, parents, the public, and students themselves that this sorting is both rational and efficient.

Assessments should not be used to channel students into dead-end mathematics courses that will leave them ill prepared to meet the challenges of the twenty-first century. The lessons of Chapter 1 remedial programs make this point clearly. Chapter 1 of Title 1 of the Elementary and Secondary Education Act provides funds for programs to improve the achievement of low-income students who are functioning below grade level in mathematics and reading. Eligibility for Chapter 1 services is determined by scores on standardized achievement tests that do not target higher-order skills. Program effectiveness is judged by performance gains on those same tests. Although the basic skills gap separating Chapter 1 students from other students has declined in the past 15 years, critics have charged that "A continued focus on remediation denies the richness of learning to those who need more, not less, of what makes education engaging and exciting."[32] In mathematics, many of the children who begin remediation in third grade never catch up to their peers. By ninth grade many are so far behind that they opt out of more challenging mathematics courses if given a choice. Others are programmed out of such courses by teachers and counselors who judge them poorly prepared for gateway subjects, such as algebra and geometry.

Many of the proposals for improving Chapter 1 programs focus on assessment, because Chapter 1 eligibility is based on students' performance on tests that reflect very low standards, and on students' relative standing. The tests typically used provide no information on how students measure up to standards of what they need to know to function in the adult world. Although proposals differ, most recommend the use of assessments consonant with the

Assessments should not be used to channel students into dead-end mathematics courses.

ideas embedded in the content and learning principles. Most also recommend that assessment results not be used for individual placements that deny children access to powerful and engaging learning experiences.[33]

PROVIDING SAFEGUARDS FOR STUDENTS

The equity principle requires that opportunity to learn be considered whenever assessment results are reported or used. Obviously, students who have had opportunities to reflect on the mathematics they are learning, to present and defend their ideas, or to organize, execute, and report on a complex piece of work will have an advantage when called upon to do so in an assessment situation. There is a legal basis for this requirement as well. The courts have held, in several landmark cases, that schools are obligated to see that all students have a chance to learn content that is assessed for high-stakes purposes.

In Florida, for example, the state legislature imposed a testing program to guarantee that all students would graduate with the minimum skills needed to function in society. To be granted a high school diploma, students had to pass functional literacy examinations in reading, writing, and mathematics in addition to completing all required courses. Average test performance was poor across the board when the program began, but minority students failed the test at disproportionately high rates. This practice was successfully opposed in the case of Debra P. v. Turlington, where the court ruled that the state "may not constitutionally so deprive its students [of a high school diploma based upon test performance] unless it has submitted proof of the curricular validity of the test which the court defined as 'a fair test of that which was taught.'" The ruling in that case and several others that dealt with high-stakes use of tests for placement established two important guidelines for fairness in educational assessment. First, assessments used to allocate different instructional treatments must be demonstrably fair tests of what students have been taught. Second, schools must show that they have taken affirmative steps to remove the effects of any past discrimination that might influence the results of the assessments.[34]

These rulings have been construed to mean that when assessments are used to make high-stakes decisions on matters such as graduation and promotion, certain basic safeguards must be built

> The courts have held that schools are obligated to provide all students a chance to learn content for high-stakes assessments.

into the system for students. Social compacts, school delivery standards, and a variety of other mechanisms have been proposed to provide the needed safeguards. The National Council on Educational Standards and Testing recommended that schools be held to standards of instructional quality (termed school delivery standards) to ensure that students have a fair chance to develop the knowledge and skills targeted by the more challenging assessments now being proposed.[35] The council was formed to advise the Congress, the Secretary of Education, and the National Education Goals Panel on the desirability and feasibility of instituting a national testing system. Other groups, while not endorsing a specific proposal for school delivery standards, have endorsed the notion that opportunity to learn must be taken into account in interpreting and making decisions based on assessment results.[36]

NSP Social Compact

We pledge to do everything in our power to ensure all students a fair shot at reaching the new performance standards, and to prevent students' performance on the new assessments from being used as the basis for awarding a diploma or any other form of credential unless all students in the jurisdiction awarding the credential have had an opportunity to prepare themselves well. This means that they will be taught a curriculum that will prepare them for the assessments, their teachers will have the preparation to enable them to teach it well, and there will be an equitable distribution of the resources the students and their teachers need to succeed.

The NSP requires the states and school districts working with them to enter into a social compact (see left) designed to address issues of fairness in accountability.[37]

Social compacts and school delivery standards are straightforward enough in conception. Both attempt to ensure that all students have a chance to learn the mathematics that is assessed. Neither proposal has found universal support among educators and policymakers, however. Challenges have been raised on technical grounds mainly related to the difficulty of developing and validating appropriate indicators and cutoff scores. Opposition has also been forthcoming on policy grounds: the burdens delivery standards might impose on states, unresolved consequences for schools that fail to satisfy standards, and unresolved questions about consequences for individual rights in high-stakes contexts when schools fail to meet delivery standards. Some of the strongest opposition to delivery standards, however, rests on the observation that remedies designed to improve schools in the past often failed precisely because the emphasis was placed on the resources schools should provide rather than the outcomes that schools should achieve.[38] The equity principle requires that due consideration be given to the consequences of assessments for individual students.

PROVIDING SAFEGUARDS FOR SCHOOLS

The equity principle also requires some consideration of assessment consequences for schools. This implies, first of all, that fair comparisons can be made of what schools are accomplishing only when assessment data include information on the nature of the students served and the adequacy of resources. Second, the principle implies that assessment results must be compiled and reported in ways that support efforts to achieve greater equity among schools. Schools educating large numbers of poor and minority students have further to go and have fewer discretionary resources with which to pursue reform than do more affluent schools. Those schools that have more to do need greater resources.

Without equity in resources—not the same as numerical equality—there can be no equity in assessment or any other sphere of education.[39] This does not absolve schools of their responsibility to educate students to the level of new standards. Rather, it obligates educational reformers and the general public to take quite strong measures to ensure that every school has the resources needed to educate its students well. Such efforts are under way in a number of states, as evidenced by recent litigation and legislative proposals in Alabama, Texas, and Massachusetts aimed at reducing disparities in the funding of poor and more affluent schools. [40]

HOLDING ALL STUDENTS TO HIGH EXPECTATIONS

Assessments can contribute to students' opportunities to learn important mathematics only if they reflect and are reinforced by high expectations for every student. Performance standards communicate expectations for achievement. They provide concrete examples and explicit definitions of what students must do to demonstrate that they have learned required content, process, and skills to a satisfactory level.[41] If we want excellence from every student, performance standards must be set high enough so that, with effort and good instruction, every student will learn important mathematics. Some students may take longer than others to master the mathematics that they will need throughout their lives. Some may require different kinds of instructional support to meet performance standards, but there can be no equity in assessment as long as excellence is not demanded of every student.

> Without equity in resources there can be no equity in assessment.

There can be no

equity in

assessment as

long as

excellence is not

demanded of

every student.

Performance standards must be high enough to ensure that doing well on an assessment means that the student has learned enough of the right kinds of mathematics to function as an informed citizen. Meeting performance standards should mean that students have an adequate foundation on which to build, regardless of their eventual chosen career.

We have much to learn about how to ensure that performance standards are uniformly high for everyone while also ensuring that students are adequately prepared to pursue their chosen field. We also have much to learn about how to maintain uniformly high performance standards while allowing for assessment approaches that are tailored to varied backgrounds. Consistent application of standards to a varied set of tasks and responses poses an enormous challenge that we do not yet know how to meet fairly and effectively. Nonetheless, the challenge is surely worth accepting.

To meet this challenge, equity must be viewed in the broadest terms possible. Equity considerations must be a part of every decision affecting the development, interpretation, and use of assessments. This view is expressed quite forcefully in a statement of equity principles adopted by a number of educational leaders, reproduced here on the following page.[42]

**Leadership Statement of Nine Principles on Equity and
Educational Testing and Assessment
March 12, 1993**

The following statement was developed by a group of educational leaders seeking to ensure that concerns for equity would be reflected in all efforts at assessment reform. Since the statement was articulated, it has been signed by educational policymakers, measurement scientists, educational researchers, school administrators, and experts on equity and diversity who represent a variety of perspectives and expertise. Although these leaders have not always agreed on the particulars of assessment reform, they have unanimously supported the need to address equity and excellence in tandem as assessment reform moves forward.

As policymakers move forward to develop new standards and assessments, they should consider including the following principles, which will help to ensure that both equity and quality are dominant themes:

1. New assessments should be field tested with the nation's diverse population in order to demonstrate that they are fair and valid and that they are suitable for policymakers to use as levers to improve outcomes before they are promoted for widespread use by American society.

2. New standards and tests should accurately reflect and represent the skills and knowledge that are needed for the purposes for which they will be used.

3. New content standards and assessments in different fields should involve a development process in which America's cultural and racial minorities are participants.

4. New policies for standards and assessments should reflect the understanding that standards and assessments represent only two of many interventions required to achieve excellence and equity in American education. Equity and excellence can only be achieved if all educators dedicate themselves to their tasks and are given the resources they need.

5. New standards and assessments should offer a variety of options in the way students are asked to demonstrate their knowledge and skills, providing a best possible opportunity for each student to perform.

6. New standards and assessments should include guidelines for intended and appropriate use of the results and a review mechanism to ensure that the guidelines are respected.

7. New policies should list the existing standards and assessments that the new standards and assessments should replace (e.g., Chapter 1 standards and tests, state-mandated student standards and tests) in order to avoid unnecessary and costly duplication and to avoid overburdening schools, teachers and students who already feel saturated by externally mandated tests.

8. New policies need to reflect the understanding by policymakers of the tradeoff between the types of standards and assessments needed for monitoring the progress of school systems and the nation versus the types of standards and assessments needed by teachers to improve teaching and learning. The attention and resources devoted to the former may compete for the limited resources available for research and development for the latter.

9. New policies to establish standards and assessments should feature teachers prominently in the development process.

ENDNOTES

[1] Edmund Gordon, "Assessment Challenges: Changing Views of Learning, Instruction and Assessment" (Paper presented at the annual conference of the National Center for Research on Evaluation, Standards, and Student Testing, Los Angeles, CA, 10 Sept. 1992).

[2] 2. National Council of Teachers of Mathematics, *Professional Standards for Teaching Mathematics* (Reston, VA: Author, March 1991), 4.

[3] *See* D. A. Archbald and F. M. Newmann, *The Nature of Authentic Academic Achievement in the Secondary School* (Reston, VA: National Association of Secondary School Principals, 1988); Joseph P. McDonald, "Three Pictures of an Exhibition: Warm, Cool, and Hard" (Substantially revised version of a paper presented at the annual meeting of the American Education Research Association, Chicago, IL, April 1991); Grant Wiggins, "Teaching to the (Authentic Test)," *Educational Leadership* 46:7 (1989), 41-47; Grant Wiggins, "Creating Tests Worth Taking," *Educational Leadership* 49:8 (1992), 26-33; Dennie P. Wolf et al., "To Use their Minds Well: Investigating New Forms of Student Assessment," in Gerald Grand, ed., *Review of Research in Education* (Washington, D.C.: American Educational Research Association, 1991), 31-74.

[4] Dennie P. Wolf and Sean F. Reardon, "Equity in the Design Performance Assessments: A Handle to Wind Up the Tongue With?" (Paper presented at the Ford Foundation Symposium on Equity and Educational Testing and Assessment, Washington, D.C., 11-12 March 1993).

[5] This exercise was created by the Balanced Assessment Project at the University of California, Berkeley, which is funded by the National Science Foundation.

[6] William L. Leap, "Assumptions and Strategies Guiding Mathematics Problem Solving by Ute Indian Students," in Rodney R. Cocking and José P. Mestre, eds., *Linguistic and Cultural Influences on Learning Mathematics* (Hillsdale, NJ: Lawrence Erlbaum Associates, Inc., 1988), 176-177.

[7] Eva L. Baker and Harold F. O'Neil, Jr., "Diversity, Assessment, and Equity in Educational Reform" (Paper presented at the Ford Foundation Symposium on Equity and Educational Testing and Assessment, Washington, D.C., 11-12 March 1993); Related papers presented at the same symposium were George F. Madaus, "A Technological and Historical Consideration of Equity Issues Associated with Proposals to Change Our Nation's Testing Policy"; Warren Simmons and Daniel P. Resnick, "National Standards, Assessment and Equity."

[8] Warren Simmons and Lauren Resnick, "Assessment as the Catalyst of School Reform," *Educational Leadership* 50:5 (1993), 11-15.

[9] "Authentic Assessments—The Challenge of Diversity and Equity," *The CRESST Line,* Newsletter of the UCLA Center for Research on Evaluation, Standards, and Student Testing, Spring 1992, 6.

[10] Robert L. Linn, Eva L. Baker, and Stephen B. Dunbar, "Complex, Performance-Based Assessment: Expectations and Validation Criteria" *Educational Researcher* 20:8 (1991), 15-21; Stephen B. Dunbar, Daniel M. Koretz, and H.D. Hoover, "Quality Control in the Development and Use of Performance Assessments," *Applied Measurement in Education* 4:4 (1991), 289-303; R.J. Shavelson, G.P. Baxter, and J. Pine, "Performance Assessments: Political

Rhetoric and Measurement Reality," *Educational Researcher* 21:4 (1992), 22-27; R.J. Shavelson et al., "New Technologies for Large-Scale Science Assessments: Instruments of Educational Reform" (Paper presented at the annual meeting of the American Educational Research Association, Chicago, IL, April 1991).

[11] Monty Neill, "Some Pre-Requisites for the Establishment of Equitable, Inclusive Multicultural Assessment Systems" (Paper presented at the Ford Foundation Symposium on Equity and Educational Testing and Assessment, Washington, D.C., 11-12 March 1993); J.R. Mercer, "Alternative Paradigms for Assessment in a Pluralistic Society," in J.A. Banks and C.A.M. Banks, eds., *Multicultural Education: Issues and Perspectives* (Boston, MA: Allyn and Bacon, 1989), 289-304; Asa G. Hilliard, III, "The Strengths and Weaknesses of Cognitive Tests for Young Children," in J.D. Andrews, ed., *One Child Indivisible* (Washington, D.C.: The National Association for the Education of Young Children, 1975), 17-33; O. Taylor and D.L. Lee, "Standardized Tests and African Americans: Communication and Language Issues," in Asa G. Hilliard, III, ed., *Testing African American Students: Special Re-Issue of the Negro Educational Review* (Morristown, NJ: Aaron Press, 1991), 67-80.

[12] David J. Clarke, "Open-Ended Tasks and Assessment: The Nettle or the Rose" (Paper presented to the Research Pre-Session of the 71st Annual Meeting of the National Council of Teachers of Mathematics, Seattle, WA, 29-30 March 1993); Stephen B. Dunbar and Elizabeth A. Witt, "Design Innovations in Measuring Mathematics Achievement" (Paper commissioned by the Mathematical Sciences Education Board, September 1993, appended to this report); *See* Patricia Ann Kenney and H. Tang, "Conceptual and Operational Aspects of Rating Student Responses to Performance Assessments" (Paper presented at the annual meeting of the American Educational Research Association, San Francisco, CA, April 1992) for a discussion of the issues involved in scoring written products in mathematics; Edward Silver and Suzanne Lane, "Balancing Considerations of Equity, Content Quality and Technical Excellence in Designing, Validating and Implementing Performance Assessments in the Context of Mathematics Instructional Reform: The Experience of the QUASAR Project (Pittsburgh, PA: Learning Research and Development Center, University of Pittsburgh, 1993).

[13] *See* Eva L. Baker, Harold F. O'Neil, and Robert L. Linn, "What Works in Alternative Assessment? (Sherman Oaks, CA: Advance Design Information, Inc., draft final report, September 1992). For discussion of broader cultural perspectives, *see* "Alternative Paradigms for Assessment in a Pluralistic Society"; "Some Pre-Requisites for the Establishment of Equitable, Inclusive Multicultural Assessment Systems"; "The Strengths and Weaknesses of Cognitive Tests for Young Children"; "Standardized Tests and African Americans: Communication and Language Issues."

[14] Lloyd Bond, "Bias in Mental Tests," in B.F. Green, ed., *New Directions for Testing and Measurement: Issues in Testing—Coaching, Disclosure and Ethnic Bias,* no. 11 (San Francisco, CA: Jossey-Bass, 1981), 55-75; Nancy Cole and Pamela Moss, "Bias in Test Use," in R. L. Linn, ed., *Educational Measurement,* 3d. ed. (New York, NY: American Council on Education/McMillan Publishing Company, 1989), 201-217.

[15] For more information on the kinds of collateral information needed to support inferences about equity from alternative assessments, *see* "Diversity, Assessment and Equity in Educational Reform"; "A Technological and

Historical Consideration of Equity Issues Associated with Proposals to Change our Nation's Testing Policy"; "To Use their Minds Well: Investigating New Forms of Student Assessment."

[16] "Diversity, Assessment and Equity in Educational Reform."

[17] Ibid.

[18] Center for the Study of Testing, Evaluation and Educational Policy, Boston College, "Progress Report: Urban District Assessment Coalition," September 1991-93; John Cawthorne, personal communication, April 1993.

[19] "Diversity, Assessment and Equity in Educational Reform"; "What Works in Alternative Assessment?"

[20] Maria Santos, personal communication, April 1993.

[21] California Assessment Program, *A Sampler of Mathematics Assessment* (Sacramento, CA: California Department of Education, 1991), 16.

[22] Richard Lesh and Susan Lamon, "Assessing Authentic Mathematical Performance," in Richard Lesh and Susan J. Lamon, eds., *Assessment of Authentic Performance in School Mathematics* (Washington, D.C.: American Association for the Advancement of Science, 1992), 35.

[23] Richard Lesh et al., "Future Directions for Mathematics Assessment," in *Assessment of Authentic Performance in School Mathematics*, 417.

[24] Ibid., 418.

[25] Ibid., 416-417.

[26] Stephen Jay Gould, *The Mismeasure of Man* (New York, NY: Norton, 1981), 146-158.

[27] M.S. Smith and J. O'Day, "Systemic School Reform," in Susan H. Fuhrman and Betty Malen, eds., *The Politics of Curriculum and Testing*, the 1990 Yearbook of Politics of Education Association (London, England: The Falmer Press, 1990), 233-267; Congressional Budget Office, *Educational Achievement: Explanations and Implications of Recent Trends* (Washington, D.C.: Government Printing Office, 1987).

[28] For discussion of issues related to sanctions, *see* Richard L. Allington and Anne McGill-Franzen, "Does High-Stakes Testing Improve School Effectiveness?" *ERS Spectrum*, 10:2 (1992), 3-12; Linda Darling-Hammond, "The Implications of Testing Policy for Quality and Equality," *Phi Delta Kappan* 73:3 (1991), 220-225; Linda Darling-Hammond, "Equity Issues in Performance-Based Assessment" (Paper prepared for the Symposium on Equity and Educational Testing and Assessment, Washington, D.C., 11-12 March 1993); Lorrie A. Shephard and Mary L. Smith, "Escalating Academic Demands in Kindergarten: Counterproductive Policies," *The Elementary School Journal* 89 (1988): 135-145; Lorrie Shepard, "Inflated Test Scores: Is the Problem Old Norms or Teaching to the Test?" *Educational Measurement Issues and Practices* 8 (1990): 15-22; Richard L. Allington and Anne McGill-Franzen, "Unintended Effects of Educational Reform in New York," *Educational Policy* 6:4 (1992), 397-414.

[29] The National Council on Education Standards and Testing, *Raising Standards for American Education: A Report to Congress, The Secretary of Education, The National Education Goals Panel, and the American People* (Washington, D.C.:

Author, 1992); M. Fleming and B. Chambers, "Teacher-Made Tests: Windows on the Classroom," in W.E. Hathaway, ed., *Testing in the Schools: New Directions for Testing and Measurement* (San Francisco, CA: Jossey Boss, 1983), 29-38; Leslie Salmon-Cox, "Teachers and Standardized Achievement Tests: What's Really Happening?" *Phi Delta Kappan* 62 (1981): 631-634; R.J. Stiggins and N.J. Bridgeford, "The Ecology of Classroom Assessment," *Journal of Educational Measurement* 22:4 (1985), 271-286; Linda Darling-Hammond and Arthur Wise, "Beyond Standardization: State Standards and School Improvement," *The Elementary School Journal* 83 (1985): 315-336.

[30] "Equity Issues in Performance-Based Assessment"; "A Technological and Historical Consideration of Equity Issues"; *Raising Standards for American Education*; National Commission on Testing and Public Policy, *From Gatekeeper to Gateway: Transforming Testing in America* (Chestnut Hill, MA: Author, 1990).

[31] Commission on Chapter 1, *Making Schools Work for Children in Poverty: A New Framework Prepared by the Commission on Chapter 1* (Washington, D.C.: American Association for Higher Education, December 1992).

[32] Ibid.; National Forum on Assessment, "Criteria For Evaluation of Student Assessment Systems, *Educational Measurement: Issues and Practice* 11:1 (1992), 32.

[33] Diana Pullin, "Legal and Ethical Issues in Mathematics Assessment" (Paper commissioned by the Mathematical Sciences Education Board, September 1993, appended to this report).

[34] *Raising Standards for American Education.*

[35] Michael T. Nettles et al., *Leadership Statement of Nine Principles on Equity and Educational Testing and Assessment*, 12 March 1993; Diversity and Equity in Assessment Network, *Guidelines for Equitable Assessment* (Cambridge, MA: Fair Test, 1993).

[36] Learning Research and Development Center, University of Pittsburgh and National Center on Education and the Economy, *New Standards Project* (Pittsburgh, PA: Author, 1993), 11.

[37] Andrew C. Porter, "School Delivery Standards," *Educational Researcher* 22:3 (1993), 24-30.

[38] "A Technological and Historical Consideration of Equity Issues Associated with Proposals to Change Our Nation's Testing Policy; "Equity Issues in Performance-Based Assessment"; "Some Pre-Requisites for the Establishment of Equitable, Inclusive Multicultural Assessment Systems."

[39] Lonnie Harp, "Texas Finance Bill Signed Into Law, Challenges Anticipated," *Education Week*, 9 June 1993; Lonnie Harp, "Impact of Texas Finance Law, Budget Increase Gauged," *Education Week*, 16 June 1993; Millicent Lawton, "Alabama Judge Sets October Deadline for Reform Remedy," *Education Week*, 23 June 1993.

[40] "National Standards, Assessment and Equity," 7.

[41] *Leadership Statement of Nine Principles on Equity and Educational Testing and Assessment.*

6 EVALUATING MATHEMATICS ASSESSMENTS

Whether a mathematics assessment comprises a system of examinations or only a single task, it should be evaluated against the educational principles of content, learning, and equity. At first glance, these educational principles may seem to be at odds with traditional technical and practical principles that have been used to evaluate the merits of tests and other assessments. In recent years, however, the measurement community has been moving toward a view of assessment that is not antithetical to the positions espoused in this volume. Rather than view the principles of content, learning, and equity as a radical break from past psychometric tradition, it is more accurate to view them as gradually evolving from earlier ideas.

Issues of how to evaluate educational assessments have often been discussed under the heading of "validity theory." Validity has been characterized as "an integrated evaluative judgment of the degree to which empirical evidence and theoretical rationales support the *adequacy* and *appropriateness of inferences* and *actions* based on test scores or other modes of assessment." [1] In other words, an assessment is not valid in and of itself; its validity depends on how it is interpreted and used. Validity is a judgment based on evidence from the assessment and on some rationale for making decisions using that evidence.

Validity is the keystone in the evaluation of an assessment. Unfortunately, it has sometimes been swept aside by other technical matters, such as reliability and objectivity. Often it has been thought of in narrow terms ("Does this assessment rank students in the same way as another one that people consider accurate?"). Today, validity is being reconceived more broadly and given greater emphasis in discussions of assessment. [2] Under this broader conception, validity theory can provide much of the technical machinery for determining whether the educational principles are met by a mathematics assessment. One can create a rough correspondence between the content principle and content validity,[3] between the learning principle and consequential or systemic validity,[4] and between the equity principle and criteria of fairness and accessibility that have been addressed by Silver and Lane.[5]

Although every mathematics assessment should meet the three principles of content, learning, and equity, that alone cannot guarantee a high-quality assessment. Technical considerations, including generalizability, evidence, and costs, still have a place. The educational principles are primary and essential but they are not sufficient.

THE CONTENT PRINCIPLE

The contexts in which assessment tasks are administered and the interpretations students make of them are critical in judging the significance of the content.

Key Questions
What is the mathematical content of the assessment?
What mathematical processes are involved in responding?

Applying the content principle to a mathematics assessment means judging how well it reflects the mathematics that is most important for students to learn. The judgments are similar to early notions of content validity that were limited to asking about the representativeness and relevance of test content. The difference lies in a greater concern today for the quality of the mathematics reflected in the assessment tasks and in the responses to them.

Procedures for evaluating the appropriateness of assessment content are well developed and widely used. Most rely heavily on expert judgment. Judges are asked how well the design of the assessment as a whole captures the content to be measured and how well the individual tasks reflect the design. The two sets of judgments determine whether the tasks sufficiently represent the intended content.

New issues arise when the content principle is applied:

• the nature of the important mathematics content leads to some types of tasks that have not been common in educational assessment,

- the emphasis on thinking processes leads to new forms of student performance, and

- the characteristics of today's important mathematics lead to a broader view of curricular relevance.

CONTENT OF TASKS

Because mathematics has been stereotyped as cut and dried, some assessment designers have assumed that creating high-quality mathematics tasks is simple and straightforward. That assumption is false. Because mathematics relies on precise reasoning, errors easily creep into the words, figures, and symbols in which assessment tasks are expressed.

Open-ended tasks can be especially difficult to design and administer because there are so many ways in which they can misrepresent what students know and can do with mathematics.[6] Students may give a minimal response that is correct but that fails to show the depth of their mathematical knowledge. They may be confused about what constitutes an adequate answer, or they may simply be reluctant to produce more than a single answer when multiple answers are called for. In an internal assessment constructed by a teacher, the administration and scoring can be adapted to take account of misunderstanding and confusion. In an external assessment, such adjustments are more difficult to make. The contexts in which assessment tasks are administered and the interpretations students are making of them are critical in judging the significance of the content.

Difficulties arise when attempts are made to put mathematics into realistic settings. The setting may be so unfamiliar that students cannot see mathematics in it. Or, the designer of the task may have strained too hard to make the mathematics applicable, ending up with an artificial reality, as in the example above.[7] As a practical matter, the angle between

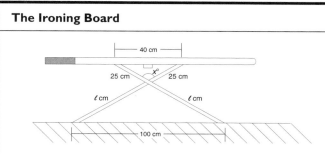

The Ironing Board

The diagram shows the side of an ironing board.
The two legs cross at x°
(a) Use the information in the diagram to calculate the angle x°.
    Give your answer to the nearest degree.
(b) Calculate the value of ℓ.

the legs of the ironing board is not nearly so important as the height of the board. As Swan notes,[8] the mathematical content is not incorrect, but mathematics is being misused in this task. A task designer who wants to claim the situation is realistic should pose a genuine question: Where should the stops be put under the board so that it will be convenient for people of different heights?

The thinking

processes

students are

expected to use

are as important

as the content of

the assessment

tasks.

The thinking processes students are expected to use in an assessment are as important as the content of the tasks. The process dimension of mathematics has not merited sufficient attention in evaluations of traditional multiple-choice tests. The key issue is whether the assessment tasks actually call for students to use the kind of intellectual processes required to demonstrate mathematical power: reasoning, problem solving, communicating, making connections, and so on. This kind of judgment becomes especially important as interesting tasks are developed that may have the veneer of mathematics but can be completed without students' ever engaging in serious mathematical thinking.

To judge the adequacy of the thinking processes used in an assessment requires methods of analyzing tasks to reflect the steps that contribute to successful performance. Researchers at the Learning Research and Development Center (LRDC) at the University of Pittsburgh and the Center for Research, Evaluation, Standards, and Student Testing (CRESST) at the University of California at Los Angeles are beginning to explore techniques for identifying the cognitive requirements of performance tasks and other kinds of open-ended assessments in hands-on science and in history.[9]

Mixing Paint

To paint a bathroom, a painter needs 2 gallons of light blue paint mixed in a proportion of 4 parts white to 3 parts blue. From a previous job, she has 1 gallon of a darker blue paint mixed in the proportion of 1 part white to 2 parts blue. How many quarts of white paint and how many quarts of blue paint (1 gallon = 4 quarts) must the painter buy to be able to mix the old and the new paint together to achieve the desired shade? How much white paint must be added and how much blue paint?

Discuss in detail how to model this problem, and then use your model to solve it.

The analysis of task demands, however, is not sufficient. The question of what processes students actually use in tackling the tasks must also be addressed. For example, could a particular problem designed to assess proportional reasoning be solved satisfactorily by using less sophisticated operations and knowledge? A problem on mixing paint, described at left, was written by a mathematics teacher to get at high-level understanding of proportions and to be approachable in a variety of ways. Does it measure what was intended?

Such questions can be answered by having experts in mathematics education and in cognitive science review tasks and evaluate student responses to provide information about the cognitive processes used. (In the mixing paint example, there are solutions to the problem that involve computation with complicated fractions more than proportional reasoning, so that a student who finds a solution has not necessarily used the cognitive processes that were intended by the task developer.) Students' responses to the task, including what they say when they think aloud as they work, can suggest what those processes might be. Students can be given part of a task to work on, and their reactions can be used to construct a picture of their thinking on the task. Students also can be interviewed after an assessment to detect what they were thinking as they worked on it. Their written work and videotapes of their activity can be used to prompt their recollections.

None of these approaches alone can convey a complete picture of the student's internal processes, but together they can help clarify the extent to which an assessment taps the kinds of mathematical thinking that designers have targeted with various tasks. Researchers are beginning to examine the structure of complex performance assessments in mathematics, but few studies have appeared so far in which labor-intensive tasks such as projects and investigations are used. Researchers at LRDC, CRESST, and elsewhere are working to develop guidelines for gauging whether appropriate cognitive skills are being engaged by an assessment task.

Innovative assessment tasks are often assumed to make greater cognitive demands on students than traditional test items do. Because possibilities for responses to alternative assessment tasks may be broader than those of traditional items, developers must work harder to specify the type of response they want to evoke from the task. For example, the QUASAR project has developed a scheme for classifying tasks that involves four dimensions: (1) cognitive processes (such as understanding and representing problems, discerning mathematical relationships, organizing information, justifying procedures, etc.); (2) mathematical content (which is in the form of categories that span the curriculum); (3) mode of representation (words, tables, graphs, symbols, etc.); and (4) task content (realistic or nonrealistic). By classifying tasks along four dimensions, the QUASAR researchers can capture much of the richness and complexity of high-level mathematical performance.

The QUASAR project has also developed a Cognitive Assessment Instrument (QCAI)[10] to gather information about the program itself and not individual students. The QCAI is a paper-and-pencil instrument for large-group administration to individual students. At each school site, several dozen tasks might be administered, but each student might receive only 8 or 9 of them. A sample task developed for use with sixth grade students is at left.[11]

The open-ended tasks used in the QCAI are in various formats. Some ask students to justify their answers; others ask students to show how they found their answers or to describe data presented to them. The tasks are tried out with samples of students and the responses are analyzed. Tasks are given internal and external reviews.[12]

Sample QUASAR Task

The table shows the cost for different bus fares.

BUSY BUS COMPANY FARES

| One Way | $1.00 |
| Weekly Pass | $9.00 |

Yvonne is trying to decide whether she should buy a weekly bus pass. On Monday, Wednesday and Friday she rides the bus to and from work. On Tuesday and Thursday she rides the bus to work, but gets a ride home with her friends.

Should Yvonne buy a weekly bus pass?
Explain your answer.

Internal reviews are iterative, so that tasks can be reviewed and modified before and after they are tried out. Tasks are reviewed to see whether the mathematics assessed is important, the wording is clear and concise, and various sources of bias are absent. Data from pilot administrations, as well as interviews with students thinking aloud or explaining their responses, contribute to the internal review. Multiple variants of a task are pilot tested as a further means of making the task statement clear and unbiased.

External reviews consist of examinations of the tasks by mathematics educators, psychometricians, and cognitive psychologists. They look at the content and processes measured, clarity and precision of language in the task and the directions, and fairness. They also look at how well the assessment as a whole represents the domain of mathematics.

The scoring rubrics are both analytic and holistic. A general scoring rubric (similar to that used in the California Assessment Program) was developed that reflected the scheme used for classifying tasks. Criteria for each of the three interrelated components of the scheme were developed at each of the five score levels from 0 to 4. A specific rubric is developed for each task, using the general scoring rubric for guidance. The process of developing the specific rubric is also iterative, with students' responses and the reactions of reviewers guiding its refinement.

Each year, before the QCAI is administered for program assessment, teachers are sent sample tasks, sample scored responses, and criteria for assigning scores that they use in discussing the assessment with their students. This helps ensure an equitable distribution of task familiarity across sites and gives students access to the performance criteria they need for an adequate demonstration of their knowledge and understanding.

CURRICULAR RELEVANCE

The mathematics in an assessment may be of high quality, but it may not be taught in school or it may touch on only a minor part of the curriculum. For some purposes that may be acceptable. An external assessment might be designed to see how students approach a novel piece of mathematics. A teacher might design an assessment to diagnose students' misconceptions about a single concept. Questions of relevance may be easy to answer.

Other purposes, however, may call for an assessment to sample the entire breadth of a mathematics curriculum, whether of a course or a student's school career. Such purposes require an evaluation of how adequately the assessment treats the depth and range of curriculum content at which it was aimed. Is each important aspect of content given the same weight in the assessment that it receives in the curriculum? Is the full extent of the curriculum content reflected in the assessment?

The term *alignment* is often used to characterize the congruence that must exist between an assessment and the curriculum. Alignment should be looked at over time and across instruments. Although a single assessment may not be well aligned with the curriculum because it is too narrowly focused, it may be part of a more comprehensive collection of assessments.

The question of alignment is complicated by the multidimensional nature of the curriculum. There is the curriculum as it exists

The term *alignment* is often used to characterize the congruence that must exist between an assessment and the curriculum.

in official documents, sometimes termed the *intended curriculum;* there is the curriculum as it is developed in the classroom by teachers through instruction, sometimes termed the *implemented curriculum;* and there is the curriculum as it is experienced by students, sometimes termed the *achieved curriculum.* Depending on the purpose of the assessment, one of these dimensions may be more important than the others in determining alignment.

Consider, for example, a curriculum domain consisting of a long list of specific, self-contained mathematical facts and skills. Consider, in addition, an assessment made up of five complex open-ended mathematics problems to which students provide multi-page answers. Each problem might be scored by a quasi-holistic rubric on each of four themes emphasized in the NCTM *Standards:* reasoning, problem solving, connections, and communication. The assessment might be linked to an assessment framework that focused primarily on those four themes.

Better methods

are needed to

judge the

alignment of new

assessments

new curricula.

An evaluator interested in the intended curriculum might examine whether and with what frequency students actually use the specific content and skills from the curriculum framework list in responding to the five problems. This examination would no doubt require a reanalysis of the students' responses because the needed information would not appear in the scoring. The assessment and the intended curriculum would appear to be fundamentally misaligned. An evaluator interested in the implemented curriculum, however, might be content with the four themes. To determine alignment, the evaluator might examine how well those themes had been reflected in the instruction and compare the emphasis they received in instruction with the students' scores.

The counting and matching procedures commonly used for checking alignment work best when both domains consist of lists or simple matrices and when the match of the lists or arrays can be counted as the proportion of items in common. Curriculum frameworks that reflect important mathematics content and skills (e.g., the NCTM *Standards* or the California Mathematics Framework) do not fit this list or matrix mode. Better methods are needed to judge the alignment of new assessments with new characterizations of curriculum.

THE LEARNING PRINCIPLE

Key Questions
How are enhanced learning and good instruction supported by the assessment?
What are its social and educational consequences?

Mathematics assessments should be judged as to how well they reflect the learning principle, with particular attention to two goals that the principle seeks to promote—improved learning and better instruction—and to its resulting goal of a high-quality educational system.

IMPROVED LEARNING

Assessments might enhance student learning in a variety of ways. Each needs careful investigation before a considered judgment is reached on the efficacy of specific assessment features. For example, a common claim is that assessment can and should raise both students' and teachers' expectations of performance, which will result in greater learning. Research on new assessments should seek to document this assertion.

Students are also presumed to need more active engagement in mathematics learning. Assessments support student learning to the extent that they succeed in engaging even those students with limited mathematical proficiency in solving meaningful problems. This support often involves activities about which students have some knowledge and interest or that otherwise motivate engagement. However, if challenging assessments are so far beyond the grasp of students whose knowledge lags behind the goals of reform, and such students are closed off from demonstrating what they do know, the assessments may well have negative effects on these students' learning. This question, like many others, deserves further investigation. In any case, student engagement in assessment tasks should be judged through various types of evidence, including teacher reports, student reports, and observations.

Learning to guide one's own learning and to evaluate one's own work is well recognized as important for developing the

> Student engagement in assessment tasks should be judged through various types of evidence, including teacher reports, student reports, and observations.

capability to continue learning. Some new forms of assessment make scoring rubrics and sample responses available to students so they can learn to evaluate for themselves how they are doing. There are indications that attention to this evaluative function in work with teachers and students has desirable effects. More research is needed to determine how best to design and use rubrics to help students' assess their own work. This is another avenue that might be explored to help assessors evaluate an assessment's potential to improve mathematics learning.

Finally, changes in student learning can be assessed directly through changes in performance over time. The nature of the assessment used to reflect change is critical. For example, should one use an assessment for which there is historical evidence, even if that assessment cannot capture changes in the mathematics considered most important for students to learn? Or should one use a new assessment reflecting the new goals but for which there is no historical evidence for comparison? The difficulty with the first situation is that it compromises the content principle. For a short time, however, it may be desirable to make limited use of assessments for which there is historical evidence and to implement, as quickly as possible, measures that better reflect new goals in a systematic way.

BETTER INSTRUCTION

Attempts to investigate the consequences of an assessment program on instruction should include attention to changes in classroom activities and instructional methods in the assignments given, in the classroom assessments used, and in the beliefs about important mathematics. Studies of the effects of standardized tests have made this point quite clearly. For example, a survey of eighth-grade teachers' perceptions of the impact of their state or district mandated testing program revealed an increased use of direct instruction and a decreased emphasis on project work and on the use of calculator or computer activities.[13] Some studies have suggested that the instructional effects of mandated testing programs on instruction have been rather limited when the stakes are low,[14] but these effects appear to increase as stakes are raised.[15] Teachers may see the effects on their instruction as positive even when those effects are directed away from the reform vision of mathematics instruction.[16]

Assessments fashioned in keeping with the learning principle should result in changes more in line with that vision. New methods

Changes in

student learning

can be assessed

directly through

changes in

performance

over time.

of assessing writing have shown how changes in instructional methods and activities can follow from reform in assessment. The change from multiple-choice tests to directed writing assessments seem to have refocused classroom instruction in California schools. A recent study showed that 90% of California teachers now assign more writing and more varied kinds of writing (e.g., narrative, persuasive).[17]

Evaluating instructional changes in mathematics requires evidence about how teachers spend their instructional time, the types of classroom activities they initiate, and how they have changed what they see as most important for instruction. Shortly after the 1989 publication of the NCTM *Standards,* a study of teachers who were familiar with the document and with its notions about important mathematics showed that they continued to teach much as they had always taught. The topics and themes recommended in the *Standards* had not been fully integrated into instruction, and traditional teaching practices continued to dominate.[18] As assessment practice changes under the guidance of the learning principle, more teaching should be in line with the reform vision, even for teachers who are not well acquainted with the *Standards* documents.

Some evidence of this change can be seen in schools where teachers are experimenting with new, more powerful forms of assessment. Early observations also raise warnings about superficial changes and about lip service paid to views that teachers have not yet internalized. Teachers weak in mathematics often have difficulty making critical judgments about the mathematics reflected in student work. They cannot differentiate confidently between correct and incorrect alternatives presented by students with novel ideas about a problem. They do not always recognize when a powerful but misconceived idea underlies an incorrect answer.[19] These observations point once again to the importance of sustained attention to the professional development of teachers. As new assessments take hold and necessary changes in curriculum and teacher development are made, the instructional effects of assessments will need to be continuously monitored and evaluated to see whether these difficulties have been overcome.

The importance of sustained attention to the professional development of teachers is critical to the success of reform.

EFFECTS ON THE EDUCATIONAL SYSTEM

Recent proposals for assessment reform and for some type of national examination system contend that new forms of assessment will promote improvements in American education. The report *Raising Standards for American Education* claims that high national standards and a system of assessment are needed because "in the absence of well-defined and demanding standards, education in the United States has gravitated toward de facto national minimum expectations."[20] The argument asserts that assessments can clarify expectations and motivate greater effort on the part of students, parents, teachers, and others involved in the educational enterprise. Evaluative questions regarding the impact of assessment on the system should concern not only the degree to which the assessments have these intended, beneficial consequences, but also the nature and size of possible unintended consequences whether positive or negative (e.g., dropout rates or tracking students).

Questions about the effects of assessment on the educational system as a whole have received increased attention in the past decade. The growing concern in measurement circles with *consequential*[21] and *systemic*[22] validity have helped guide the discourse. Consequential validity refers to the social consequences that the use of an assessment can have. For example, teachers may adjust their instruction to reflect assessment content. They may spend class time using practice materials that match the assessment. Evidence needs to be collected on the intended and the unintended effects of an assessment on how teachers and students use their time and conceive of their goals.[23] Systemic validity refers to the curricular and instructional changes induced in the educational system by an assessment. Evaluating systemic effects thoroughly is a massive undertaking, and there are few extant examples in assessment practice. Even so, it is important to keep a systemic orientation in mind, for the potential impact of assessment on instruction and learning can not be separated from broader educational considerations.

Curricular Effects A comprehensive evaluation of the consequences of any assessment system would include evidence about the impact of the assessments on the curriculum. Influences on curriculum include changes in the way instructional time is allocated and in the nature of the assignments given students. Evidence of curriculum changes can be obtained through questionnaires given to teachers or students, logs kept by teachers on actual class activities, or observations of classrooms. Most of the research on curriculum changes as a consequence of mandated testing programs has made use of teacher questionnaires.[24]

The intended and unintended effects of an assessment on how teachers and students use their time and conceive of their goals should be studied.

Outside Effects Assessments such as the Scholastic Assessment Test (SAT) and the Advanced Placement (AP) test are undergoing fundamental change with widespread impact. The use of calculators on the calculus AP exam, for example, is having a profound effect on many high school teachers and way they use technology in their classrooms.[25]

Another aspect of system effects sought in many reform efforts is to change the attitudes of parents, policymakers, and other citizens about the nature of mathematics students need to learn and to bring each group into a closer and more supportive role with the efforts of the school. Although there is little evidence to date on these systemic effects, it will be important to evaluate any change in such attitudes and actions which may evolve with changes in assessment.

· ·

THE EQUITY PRINCIPLE

Key Questions
Does the assessment favor one group over others for reasons irrelevant to what it is intended to measure?
How justifiable are comparisons with a standard or with the performance of others?
Are the tasks accessible to these students?

Several aspects of the principle require examination and evaluation. The first aspect involves the usual issues associated with equity of assessment: traditional questions of fairness and of comparability across groups, scores, tasks, and situations. The second aspect involves questions of whether students have had the opportunity to learn important mathematics (whether they have been taught the important mathematics being assessed). The third aspect is newer and is associated with pedagogy that requires that all students find assessment tasks accessible if the tasks are to have the needed positive impact on their learning.

FAIRNESS AND COMPARABILITY

Traditional concerns with fair assessment are amplified or take on different importance in the context of new forms of mathematics assessment. For example, when an assessment includes a few complex tasks, often set in contexts not equally familiar to all students, any systematic disparity in the way tasks are devised or chosen becomes magnified. These disparities become critical if group differences in performance are due primarily to differences in familiarity with the contexts rather than to the underlying mathematical skills and concepts. Systematic differences in scorers' judgments may exacerbate the problem.

Equity challenges are raised by new emphases in mathematics assessments on types of student performance, such as communication, that once were considered quite separate from mathematics. As researchers using a variety of assessment formats are learning, tasks that require explanation, justification, and written reflection may leave unclear the extent to which student performance reflects different knowledge of mathematics and different language and communication skills. [26] Even though both are important, inferences about groups may be misleading if distinctions between these different skills are not made. [27]

Assessors need to know how far they are justified in comparing mathematics performance across sites (e.g., school to school, state to state) and across levels (e.g., school to district, student to state or national norm). Comparisons also need to be justified if they are going to be made from year to year. Such comparisons can be examined either statistically or by a judgmental-linking process.[28] In either case, the adequacy of the comparison for different groups of students should be demonstrated.

OPPORTUNITY TO LEARN

Students have different learning opportunities at home and at school, and such opportunities can be a major factor in their performance on educational assessments. Differences in educational experiences between groups have, for example, been a major explanatory variable in studies of group differences. However, in an educational setting designed to provide all students the opportunity to learn important mathematics, such differences may take on other implications. For reasons of equity, all students need to have experienced the important mathematics being assessed. Thus, studies of the congruence of assessments with the instructional experiences of all students are needed.

Different learning opportunities at home and at school can be a major factor in students' performance on educational assessments.

When the assessments have high-stakes implications for students, demonstrating opportunity to learn is essential. For example, in court cases involving state assessments required for high school graduation, it has been incumbent on the state to demonstrate that the skills tested are a part of the curriculum of all students, that students had an opportunity to learn the material tested. When educational reformers propose high-stakes assessment of important subject matter as a basis for key school-leaving certifications, similar opportunity-to-learn questions are raised.

ACCESS

Assessors need to know whether an assessment has offered all students the opportunity to engage in mathematical thinking. Today's vision of mathematics instruction requires a pedagogy that helps students learn mathematics that is accessible to them and relevant to their lives. This requirement affects assessment as well. If students are to find mathematics approachable, assessments must provide ways for every student to begin work on problems even if they cannot complete them as fully or as well as we would wish. Assessments can be evaluated according to how well they provide such access.

Traditionally, it has been recognized that assessment as a whole must be in the range of what a target group can do, or little information about the group tested is gained from the assessment. Consequently a range of difficulty is built into most assessments with the expectation that the majority of students will be able to complete the easiest tasks and only a few the most difficult ones. Typically, the target level of difficulty for tests has been set so that the test will seem fairly difficult to students, with slightly more than 50 percent of the questions answered correctly on average.

When the tasks are more complex and there are fewer of them, perceived difficulty of tasks takes on greater significance. If the goal is to help support students' opportunity to learn important mathematics, perceived difficulty must be taken into account along with other aspects of accessibility. The role assessments play in giving students the sense that mathematics is something they can successfully learn is largely

The role assessments play in giving students the sense that mathematics is something they can successfully learn is largely unexplored territory.

unexplored territory. That territory needs to be explored if mathematics assessments are to be evaluated with respect to the equity principle.

. . . .

GENERALIZATION

Key Questions
What generalizations can be made about student performance? How far are they justified?

A major issue in using alternative assessments concerns the inferences that can be drawn from them. To what extent can an assessor generalize from a student's score on this particular set of tasks to other tasks, on the format of tasks to other formats, on the occasion of assessment to other occasions, on the particular scorers to other scorers? Accumulating evidence suggests that in the areas explored most extensively to date (writing and hands-on science assessment), relatively good reliability of the assessment process can be obtained by systematically training the raters, generating performance tasks systematically, using explicit criteria for scoring responses, and using well-chosen sample responses to serve as anchors or exemplars in assigning performance levels.[29] Although few innovative mathematics assessments have been in place long enough to provide a solid base from which to draw conclusions, recent research suggests that acceptable levels of consistency across raters may be achievable in mathematics as well.[30]

A second aspect of generalizability reflects whether the alternative assessment measures the particular set of skills and abilities of interest within a domain. This aspect may represent a special challenge in mathematics, particularly as assessments strive to meet broader goals.[31] As researchers using a variety of assessment formats are discovering, tasks that require explanation, justification, and written reflection leave unclear the extent to which student performance reflects knowledge of mathematics rather than language and communication skills.[32]

A third aspect of generalizability rests on the consistency of scores over different tasks which can be thought of as task comparability. High levels of task generalizability are required to draw broad inferences about a learner's mathematical development or compe-

A broad variety of tasks are needed to measure the facets of proficiency and performance that make up mathematical competence.

tence from performance on a specific set of tasks. Research in mathematics is beginning to replicate the central finding from investigations of performance assessments in other content areas: The more direct methods of assessing complex performance do not typically generalize from one task to another.[33] It may be necessary to administer a broad variety of tasks to measure the many facets of proficiency and performance that make up mathematical competence. It will be essential to continue evaluating evidence on generalizability as new forms of assessment are widely used.

- -

EVIDENCE

Key Questions
What evidence does the assessment provide?
What is the value of that evidence?

Current interest in broadening the range of mathematics assessment tasks reflects a desire that students engage in important mathematical thinking: creative, motivated, associative thinking that is a goal of mathematics education. Whereas traditional item-writing procedures and test theory focus attention on the measurement properties of an assessment, the content, learning, and equity principles recognize the educational value of assessment tasks. However, if inferences and decisions are to be made about school systems or individual students, educational values cannot be the only ones present in the analysis. Especially as stakes increase, the assessor must ensure that the evidence an assessment evokes and the way it is interpreted can be justified, possibly in a court of law. The nature, weight, and coverage of information an assessment provides for a given purpose determines its value as evidence of student learning.

Assessment data, such as right or wrong responses, methods of solution, or explanations of approach, are not inherently evidence in and of themselves. They become evidence only in relation to various inferences made.[34] Moreover, a given observation can provide direct evidence about certain inferences and indirect evidence about others. It may provide conclusive evidence about some inferences, moderate evidence about others, and none whatsoever about still others. The central question, then, is *Evidence about what?* Until it has been answered, *How much evidence?* cannot

The nature, weight, and coverage of information an assessment provides for a given purpose determines its value as evidence of student learning.

even be posed. The best guideline is more of a meta-guideline: First determine what information is needed, and then gauge the effectiveness and efficiency of an assessment in providing such information.

Different kinds and amounts of evidence must be gathered to suit different purposes. For example

Determine what information is needed and then gauge the effectiveness and efficiency of an assessment in providing such information.

- Do inferences concern a student's comparative standing in a group of students, that is, are they *norm-referenced*, or do they gauge the student's competencies in terms of particular levels of skills or performance, that is, are they *criterion referenced*? A norm-referenced assessment assembles tasks to focus evidence for questions such as, *Is one student more or less skilled than another?* Accuracy of comparisons is of interest, which is reflected by traditional reliability coefficients. A criterion-referenced assessment in the same subject area might have tasks that were similar but selected to focus evidence for questions such as, *What levels of skills has this student attained?* Reliability coefficients are irrelevant for this purpose; what matters is the weight of evidence for the inference, to be established by investigating how such inferences vary with more or fewer tasks of various types.

- Will important decisions be based on the results, that is, is it a high-stakes assessment? A quiz to help students decide what to work on today is a low-stakes assessment; a poor choice is easily remedied. An assessment to determine whether they should graduate from eighth grade is high stakes at the level of the individual. An assessment to distribute state funds among schools is high stakes at the level of the school. Any assessment that supports decisions of consequence must provide commensurately dependable evidence. For high-stakes decisions about individuals, for example, several tasks may be necessary to establish the range and level of each student's proficiency.

- What is the relationship between tasks and the instructional backgrounds of students? When the purpose of assessment is to examine students' competence with the concepts they have been studying, an assessment is built around tasks that are relevant to those concepts in various ways. Because the students have been provided requisite instruction, extended and focused tasks can be quite informative. If the assessment is designed to survey a broad range of proficiency across the state, it is unconscionable for students whose classroom experience has left them unprepared to spend two days on a task that is inaccessible to them.

- Do inferences concern the competencies of individual students, as with medical certification examinations, or the distributions of competencies in groups of students, as with the National Assessment of Educational Progress (NAEP)? When the focus is on the individual, enough evidence must be gathered about each student to support inferences about the student specifically. On the other hand, a bit of information about each of several students—too little to say much about any of them as individuals—can suffice in the aggregate to monitor the level of performance in a school or a state. In these applications, classical reliability coefficients can be abysmal but that is not relevant. The pertinent question is whether accuracy for group characteristics is satisfactory.

- To what degree can and should contextual information be taken into account in interpreting assessment results? Mathematics assessment tasks can be made more valid if they broadly reflect the range of mathematical activities people carry out in the real world. This includes features not traditionally seen in assessments: collaborative work, variable amounts of time, or resources such as spreadsheets or outside help. The classroom teacher is in a position to evaluate how these factors influence performance and effectively take it into account in inferences about the students' accomplishments and capabilities. It is not so easy for the state director of assessment or the chief state school officer who must deal with results from a quarter million students to appreciate when students' responses reflect more or less advantaged circumstances.

Just looking at an assessment task or a collection of tasks cannot indicate whether the assessment will serve well for a given purpose. The tasks provide an opportunity to gather evidence; whether it is acceptable for a given use depends critically on what that use will be. High-stakes decisions for individuals are most demanding in the sense that they require strong enough evidence about each and every individual about whom decisions are being made to justify those decisions: to the student, to the parent, and, increasingly often, to the court. The same amount of time on the same tasks found inadequate for a high-stakes decision about individual students, however, may be quite satisfactory for high-stakes decisions about schools or for low-stakes instructional feedback to individual students.

COSTS AND BENEFITS

Key Questions
What are the costs of the assessment?
What are the benefits?

In traditional educational testing, the guidelines for evaluation of assessment tasks concerned almost exclusively how consistently and how well they ordered individual students along a scale. This view shaped the evolution of testing to favor multiple-choice tasks because they were the most economical, within a traditional cost/benefits framework. However, if one is interested in a expanded range of inferences about student learning, or if one takes a broader view of the potential values of assessment tasks, then the cost/benefits equation is changed. Whenever decisions of consequence are to be made from assessment results, it is incumbent on the assessor to characterize the evidence from the assessment tasks on which the decision is based.

Assessments must be feasible. They need to be practical and affordable, credible to the profession, and acceptable to the public. The following estimates have been offered for the development, administration, and scoring costs of different assessments in use today:[35]

- Commercial standardized test: $2 to $5 per student

- NAEP (1 hour, machine scorable): $100 per student

- European experience (essay exams of four to six questions): $135 per student

- AP exams: $65 per subject or $325 per student for the five-battery test proposed by the National Council on Education Standards and Testing

- Estimated total cost for AP-type exam, three grade levels per year: $3 billion annually

A recent study by the General Accounting Office (GAO)[36] of costs of a national examination yielded much lower estimates for performance assessments:

- Systemwide multiple-choice tests in four or five subjects (including staff time): $15 per student

- Systemwide performance-based tests used in some states (including staff time): $33 per student

- Estimated total cost for a national test modeled on systemwide multiple-choice tests: $160 million annually

- Estimated total cost for a national test modeled on system-wide performance-based tests: $330 million annually.

Although the earlier estimate of $325 per student annually was undoubtedly inflated because it did not take into account some of the savings that might be realized in a national examination if it were not based on the AP model, the GAO estimate of $33 seems very low.[37] The GAO survey oversampled seven states that were using performance-based formats in state-mandated testing. Two states that were experimenting with portfolio assessments, Arizona and Vermont, felt that portfolio assessments were not "tests," and, as a result, did not complete that portion of the survey.[38] Something closer to the European figure of $135 per student may be more plausible than $33.

Whatever the estimate, performance assessment in mathematics is clearly going to be more expensive than standardized multiple-choice tests have been. Standardized tests have costs that are clearly defined. Such tests may be very costly to develop, but the costs can be amortized over millions of students taking the test over several years. Performance assessment brings high development costs together with additional costs of training teachers to

Assessments need to be practical and affordable, credible to the profession, and acceptable to the public.

administer the assessment and paying for scoring. These costs are often hard to detect because local districts pay the cost of substitutes for the teachers who are being trained or doing scoring.[39]

Performance assessments can take time that might be used for other instruction. By one estimate,[40] the Standard Assessment Tasks recently introduced in Great Britain and scheduled to take 3 weeks were estimated by local administrators to require closer to 6 weeks. In Frederick County, Maryland, classes in some grades lost a whole week of instruction completing performance assessments in mathematics and language arts.[41]

These estimates of direct costs may understate the benefits of performance assessments because innovative assessments contribute to instruction and teacher development. Thus, significant portions of assessment could be "charged" to other accounts.[42] As noted elsewhere in this report, the benefits of good assessments are many. Time spent on high-quality mathematics assessment is time *well* spent because such assessment contributes directly to the learning process.

Assessment, even performance assessment, can be made relatively affordable, as experience with new examinations in various European countries suggests. The problem may be that when the same assessment is used for instructional purposes and accountability purposes, the price gets inflated. If assessment contributes to teaching and learning, then a major cost (administration time) can be attributed to instruction,[43] since time spent looking at students' work or listening to their reports is time the teacher may need to spend as part of instruction anyway. It is the external markers, the monitoring schemes, and the policing of teacher judgments that impose the true added costs.

Educators appreciate the need for a broad view of the goals of assessment and for what constitutes good evidence that the goals are being met. Assessors need more evidence on matters such as the effects of administering one assessment rather than another. More importantly from the mathematics teacher's point of view, if the mathematics assessed is not good mathematics that relates to the student's learning, all the validity coefficients in the world will be of little value.

Time spent on high-quality mathematics assessment is time *well* spent because such assessment contributes directly to the learning process.

POSTSCRIPT

The three principles proposed here—content, learning, and equity—function as a gyroscope for reform of mathematics assessment, one that will help keep mathematics reform on course to new assessments that reflect important mathematics, support good instruction, and promote every student's opportunity to learn.

The guidance system of this gyroscope provides a powerful tool in the journey toward assessment reform. However, it is only a tool, not in itself sufficient to the task. Equally important is the worthiness of the vessel for the voyage, a crew capable of making necessary midcourse corrections, and a detailed navigation chart showing the desired port.

The vessel of reform is the nationwide focus on systemic change: a coordinated response of all major components of the educational system (curriculum, teaching, assessment, governance, teacher education, school organization, etc.). In mathematics, the vessel is particularly sturdy and well launched on its journey. Already available are descriptions of the challenge (*Everybody Counts*), goals for what students should learn (*Curriculum and Evaluation Standards*), and teaching methods needed in support of that learning (*Professional Standards for Teaching Mathematics*). NCTM is now developing a third in its series of standards volumes, this one on assessment. Scheduled for release in spring 1995, this volume will lay out standards for assessments that serve a range of purposes from classroom instruction to policy, program evaluation, planning, and student placement. The three components of standards—curriculum, pedagogy, and assessment—provide a basis for renewing teacher education, rethinking school organization, enhancing implementation of reform, and promoting dialogue about systemic change among the many stakeholders in mathematics education.

Provisions for the voyage are supplied by material resources that stimulate wide-spread participation in assessment reform. These resources provide a rich array of examples of high-quality assessment consonant with the vision of mathematics and mathematics education expressed in the *Standards*. Some provide specific examples to exemplify overarching ideas (e.g., *Mathematics Assess-*

ment: *Myths, Models, Good Questions, and Practical Suggestions*[44] and *Measuring Up*). Others rely on specific examples that emerge from large-scale projects with schools nationwide, such as the New Standards Project, QUASAR, and the materials that will emerge from projects supported by the National Science Foundation and other funding sources. *Measuring What Counts* enhances this suite of resources by providing a conceptual guide to move states, districts, and individuals ahead in their thinking about assessment reform.

All educational

actions must

support

students'

learning of more

and better

mathematics;

assessment

is no exception.

Individuals from all parts of the educational system bring different talents and insights to their role as crew on the voyage of assessment reform. Teachers are the captains, charged with the front-line responsibility of providing high-quality mathematics education to all students. Many in the measurement community are exploring new paradigms consonant with principles of validity, reliability, generalizability, and other psychometric constructs. Teacher educators see in innovative assessment the opportunity and necessity to enrich both teacher preparation and professional development. New assessments encapsulate what is valued in mathematics education and often provide the basis for creating a shared vocabulary about the needed changes among faculty. Content specialists are exploring the use of assessment as a lever to create significant curricular and pedagogical change, making "teaching to the test" a positive force for change. Researchers in mathematics education are examining many unresolved questions about how cognitive, affective, and social factors relate to students' performance on assessments. Assessment researchers are rethinking basic measurement constructs and refining their tools to be appropriate both to the kinds of assessments now favored by educators and to the new functions that assessment is expected to serve, as a guidance system for educational reform.[45] Policymakers are speaking out on behalf of systemic change, with a deep understanding of the potential for new assessments to move the entire enterprise forward.

Many organizations are emerging on local, state, and national levels to broaden the recruitment of new members. Networks and alliances such as State Coalitions for Mathematics and Science Education, the Alliance to Improve Mathematics for Minorities, the State Systemic Initiatives, and the Math Connection are defining their mission to promote reform in mathematics education, including assessment that meets the content, learning, and equity principles.

Through these organizations, people are finding new ways to communicate and to explore new ideas. For example, an emerging network linking measurement and mathematics content experts will help promote development of high-quality assessment instruments. This rich flow of information helps keep reform on course as more is learned about potential trouble spots and potential solutions become quickly and widely disseminated.

The destination for the voyage of reform is well-known: every student must learn more mathematics. All educational actions must support this goal, and assessment is no exception. Although there are many unanswered questions that will require continuing research, the best way for assessment to support the goal is to adhere to the content, learning, and equity principles.

ENDNOTES

[1] Samuel Messick, "Validity," in R.L. Linn, ed., *Educational Measurement* (New York, NY: American Council on Education/Macmillan, 1989), 13.

[2] See *Educational Measurement*; Robert L. Linn, "Educational Assessment: Expanded Expectations and Challenges," *Educational Evaluation and Policy* 15:1 (1993), 1-16; Robert L. Linn, Eva L. Baker, and Stephen B. Dunbar, "Complex, Performance-Based Assessment: Expectations and Validation Criteria," *Educational Researcher* 20:8 (1991), 15-21.

[3] "Validity."

[4] "Complex, Performance-Based Assessment"; John R. Frederiksen and Allan Collins, "A Systems Approach to Educational Testing," *Educational Researcher* 18:9 (1989), 27-32.

[5] Edward Silver and Suzanne Lane (Remarks made at the Ford Foundation Symposium on Equity and Educational Testing and Assessment Washington, D.C., 11-12 March 1993).

[6] David J. Clarke, "Open-Ended Tasks and Assessment: The Nettle or the Rose" (Paper presented to the research pre-session of the 71st annual meeting of the National Council of Teachers of Mathematics, Seattle, WA, 29-30 April 1993).

[7] Malcolm Swan, "Improving the Design and Balance of Assessment," in Mogens Niss, ed., *Investigations into Assessment in Mathematics Education: An ICMI Study* (Dordrecht, The Netherlands: Kluwer Academic Publishers, 1993), 212.

[8] Ibid.

[9] See for example, Eva L. Baker, Harold F. O'Neil, Jr., and Robert L. Linn, "Policy and Validity Prospects for Performance-Based Assessment" (Paper presented at the annual meeting of the American Psychological Association, San Francisco, CA, August 1991); Robert Glaser, Kalyani Raghavan, and Gail Baxter, *Cognitive Theory as the Basis for Design of Innovative Assessment* (Los Angeles, CA: The Center for Research on Evaluation, Standards, and Student Testing, 1993).

[10] Edward A. Silver and Suzanne Lane, "Assessment in the Context of Mathematics Instruction Reform: The Design of Assessment in the QUASAR Project," in Mogens Niss, ed., *Cases of Assessment in Mathematics Education: An ICMI Study* (Dordrecht, The Netherlands: Kluwer Academic Publishers, 1993), 59-69.

[11] Edward A. Silver and Suzanne Lane, "Balancing Considerations of Equity, Content Quality, and Technical Excellence in Designing, Validating and Implementing Performance Assessments in the Context of Mathematics Instructional Reform: The Experience of the QUASAR Project" (Pittsburgh, PA: Learning Research and Development Center, University of Pittsburgh, Draft version, February 1993).

[12] Ibid., 47.

[13] Thomas A. Romberg, E. Anne Zarinnia, and Steven R. Williams, "Mandated School Mathematics Testing in the United States: A Survey of State Mathematics Supervisors" (Madison, WI: National Center for Research in Mathematical Sciences Education, September 1989).

[14] D.R. Glasnapp, J.P. Poggio, and M.D. Miller, "Impact of a 'Low Stakes' State Minimum Competence Testing Program on Policy, Attitudes, and Achievement," in R.E. Stake and R.G. O'Sullivan, eds., *Advances in Program Evaluation: Effects of Mandated Assessment on Teaching*, vol. 1, pt. b (Greenwich, CT: JAI Press, 1991), 101-140.

[15] H. D. Corbett and B.L. Wilson, *Testing, Reform, and Rebellion* (Norwood, NJ: Ablex, 1991).

[16] Lynn Hancock and Jeremy Kilpatrick, *Effects of Mandated Testing on Instruction* (Paper commissioned by the Mathematical Sciences Education Board, September 1993, appended to this report).

[17] John O'Neil, "Putting Performance Assessment to the Test," *Educational Leadership* 49:8 (1992), 14-19; Joan L. Herman, "What Research Tells Us About Good Assessment," *Educational Leadership* 49:8 (1992), 74-78.

[18] Iris R. Weiss, Jan Upton, and Barbara Nelson, *The Road to Reform in Mathematics Education: How Far Have We Traveled?* (Reston, VA: National Council of Teachers of Mathematics, 1992).

[19] Dennie P. Wolf, "Assessment as an Episode of Learning," *Taking Full Measure: Rethinking Assessment Through the Arts* (New York, NY: College Entrance Examination Board, 1991), 57; Thomas A. Romberg, "What We Know: Small Group Sessions-Math" (Presentation made at the National Center for Research on Evaluation, Standards, and Student Testing, Los Angeles, CA, 10-11, September 1992); Meryl Gearhart et al., "What We Know: Small Group Sessions-Portfolios" (Presentation made at the National Center for Research on Evaluation, Standards, and Student Testing, Los Angeles, CA, 10-11, September 1992).

[20] The National Council on Education Standards and Testing, *Raising Standards for American Education: A Report to Congress, The Secretary of Education, The National Education Goals Panel, and the American People* (Washington, D.C.: Author, 1992), 2.

[21] "Validity," 13.

[22] "A Systems Approach to Educational Testing."

[23] "Complex, Performance-Based Assessment."

[24] *Effects of Mandated Testing on Instruction*; "Impact of a 'Low Stakes' State Minimum Competency Testing Program on Policy, Attitudes, and Achievement"; George F. Madaus et al., *The Influence of Teaching Math and Science in Grades 4-12. Executive Summary* (Partial results of a study conducted by the Center for the Study of Testing, Evaluation and Educational Policy, 1992).

[25] Franklin Demana and Bert Waits, "Implementing the Standards: The Role of Technology in Teaching Mathematics," *Mathematics Teacher* 83:1 (1990), 27-31.

[26] Edward A. Silver, Patricia Ann Kenney, and Leslie Salmon-Cox, *The Content and Curricular Validity of the 1990 NAEP Mathematics Items: A Retrospective Analysis* (Pittsburgh, PA: Learning Research and Development Center, University of Pittsburgh, 1991), 25; "Design Innovations in Measuring Mathematics Achievement"; Dennie Wolf, session on "What Can Alternative Assessment Really Do for Us?" (Presentation made at the National Center for Research on Evaluation, Standards, and Student Testing, Los Angeles, CA, 10-12 September 1992).

[27] Student unfamiliarity with the required mode of response is frequently cited as a potential source of bias. One study showed that even when an instructional intervention was used to provoke multiple responses from students, there was no corresponding increase in the level of their responses. (P. Sullivan, D. J. Clarke, and M. Wallbridge, *Problem Solving with Conventional Mathematics Content: Responses of Pupils to Open Mathematical Tasks*, Research Report 1 (Oakleigh, Australia: Mathematics Teaching and Learning Centre, Australian Catholic University, 1991). Thus, it may take more than familiarity to overcome the problem.

[28] The categories come from Robert J. Mislevy, *Linking Educational Assessments: Concepts, Issues, Methods, and Prospects* (Princeton, NJ: Educational Testing Service, Policy Information Center, 1992).

[29] Stephen B. Dunbar, Daniel M. Koretz, and H. D. Hoover, "Quality Control in the Development and Use of Performance Assessments" *Applied Measurement in Education* 4:4 (1991), 289-303; Joan Herman, *What's Happening with Educational Assessment?* (Los Angeles: Co-published by UCLA CRESST and SouthEastern Regional Vision for Education (SERVE), June 1992); U.S. Congress, Office of Technology Assessment, *Testing in American Schools: Asking the Right Questions,* OTA-SET-519 (Washington, D.C.: U.S. Government Printing Office, 1992).

[30] Suzanne Lane et al., "Reliability and Validity of a Mathematics Performance Assessment," *International Journal of Educational Research,* in press.

[31] *Design Innovations in Measuring Mathematics Achievement.*

[32] *The Content and Curricular Validity of the 1990 NAEP Mathematics Items: A Retrospective Analysis;* Dennie Wolf (Remarks made at the National Center for Research on Evaluation, Standards, and Student Testing, Los Angeles, CA, 10-12 September 1992).

[33] "Quality Control in the Development and Use of Performance Assessments"; *What's Happening with Educational Assessment?*; Gail Baxter et al., "Mathematics Performance Assessment: Technical Quality and Diverse Student Impact," *Journal for Research in Mathematics Education* 24:3 (1993), 190-216. "Complex, Performance-Based Assessment"; Richard Shavelson, Gail Baxter, J. Pine, "Performance Assessment: Political Rhetoric and Measurement Reality, *Educational Research* 21:4, (1992), 22-27; Richard Shavelson et al., "New Technologies for Large-Scale Science Assessments: Instruments of Educational Reform" (Paper presented at the annual meeting of the American Educational Research Association, Chicago, IL, 1991); *The Content and Curricular Validity of the 1990 NAEP Mathematics Items: A Retrospective Analysis;* M. A. Ruiz-Primo, Gail Baxter, and Richard Shavelson, "On the Stability of Performance Assessments," *Journal of Educational Measurement* 30:1 (1993), 41-53 found moderate generalizability across occasions for a hands-on science investigation and a notebook surrogate, and the procedures students used tended to change across occasions.

[34] David A. Schum, *Evidence in Inference for the Intelligence Analyst* (Lanham, MD: University Press of America, 1987), 16.

[35] Koretz et al., Testimony before Subcommittee on Elementary, Secondary, and Vocational Education (Committee on Education and Labor, U.S. House of Representatives, 19 February 1992); General Accounting Office, *Student Testing: Current Extent and Expenditures, with Cost Estimates for a National*

Examination, Report to Congressional Requesters, GAO/PEMD-93-8 (Gaithersburg: Author, January 1993). Note that the cost figure for NAEP is misleading: NAEP is not all machine scorable. Furthermore, NAEP uses a sample of students to make inferences about the population of students at a grade level, so the total cost is less than that of administering a less expensive test to the entire population.

[36] *Student Testing: Current Extent and Expenditures with Cost Estimates for a National Examination.*

[37] Daniel M. Koretz, personal communication, 29 June 1993.

[38] *Student Testing*, 15.

[39] Ibid.

[40] Testimony Before the Subcommittee on Elementary, Secondary, and Vocational Education.

[41] Pamela Aschbacher, "What We Know: Small Group Sessions-Multidisciplinary" (Presentation made at the National Center for Research on Evaluation, Standards, and Student Testing, 10-12 September 1992).

[42] Ruth Mitchell, *Testing for Learning: How New Approaches to Evaluation Can Improve American Schools* (New York, NY: The Free Press, 1991); Alan Bell, Hugh Burkhardt, and Malcolm Swan, *Assessment of Authentic Performance in School Mathematics* (Washington, DC: American Association for the Advancement of Science, 1992).

[43] *Assessment of Authentic Performance in School Mathematics.*

[44] The National Council of Teachers of Mathematics, Inc., *Mathematics Assessment: Myths, Models, Good Questions, and Practical Suggestions*, Jean Kerr Stenmark, ed., (Reston, VA: The National Council of Teachers of Mathematics, Inc., 1991).

[45] Lee J. Cronbach, "Five Perspectives on Validity Argument," in Howard Warner and Henry I. Baum, *Test Validity* (Hillsdale, NJ: Lawrence Erlbaum Associates, Inc., 1988); "Complex, Performance-Based Assessment"; "Validity"; Pamela Moss, "Shifting Conceptions of Validity in Educational Measurement" (Paper presented at the annual meeting of American Educational Research Association, San Francisco, April 1992).

COMMISSIONED PAPERS

EFFECTS OF MANDATED TESTING ON INSTRUCTION

DESIGN INNOVATIONS IN MEASURING MATHEMATICS
ACHIEVEMENT

LEGAL AND ETHICAL ISSUES IN MATHEMATICS
ASSESSMENT

Effects of Mandated Testing on Instruction

LYNN HANCOCK
JEREMY KILPATRICK
UNIVERSITY OF GEORGIA

The past two decades have seen a striking increase in the use of testing in the United States by school officials and legislators attempting to determine whether funds invested in schools are yielding an educated citizenry. Testing is viewed as the major instrument for holding schools accountable for the resources they have received. It has become a vital tool of state and federal education policy. Governments and local school authorities have mandated the administration of tests, usually at the end of major phases of schooling but sometimes at the end of each grade, in the belief that test scores provide critical information on how well students are learning and how effective instruction has been.

Testing of all types seems to be on the rise in the United States, but the increase in mandated testing has been especially dramatic. In 1990, 46 states had mandated testing programs as compared with 29 in 1980. As the school population increased 15 percent from 1960 to 1989, revenues from the sales of standardized tests increased 10 times as fast.[1] More than a third of the elementary school teachers in a recent survey[2] saw the emphasis on standardized testing in U.S. education as strong and getting stronger. Somewhat fewer saw the same increasing emphasis in their school district, and even fewer saw it in their own school. Almost no teachers, however, said that the emphasis on standardized testing was weak or diminishing.

As the amount of instructional time lost to mandated testing increases, teachers and other educators have begun to express concern about the effects of such testing on instruction. Because most of the standardized tests used in mandated testing programs are of the multiple-choice variety, particular attention has been given to the argument that these tests promote a narrow approach to teaching, passive and low-level forms of learning, and a fragmented school curriculum.

The amount of available research to address these concerns and arguments, however, is quite sparse. Much of this research consists of surveys of teachers' appraisals of the effects of mandated testing rather than direct observation or independent judgments of these effects. The findings from the research are often inconclusive and sometimes conflicting.

The purpose of this paper is to review the literature on the effects of mandated testing on school instruction. Because the climate of educational testing in the United States has changed so rapidly over the past decade, we give special attention to the most recent studies. Furthermore, although the research is not confined to mandated testing in mathematics, we have tried to draw conclusions of particular relevance to the mathematics education community.

EFFECTS ON CURRICULUM

Resnick and Resnick[3] portray the process by which state legislatures and departments of education use accountability programs to control curriculum content and standards of performance as follows. Tests of desired educational objectives are mandated and administered, and the scores are widely disseminated. Because of the attention given to test results, teachers gradually adapt their instruction to the test objectives and format. Adaptation of the curriculum takes place as teachers who administer a test every year have the opportunity to see test forms and compare test content with the content they are teaching. The result is that "you get what you assess, and you do not get what you do not assess."[4]

Others have also described this process. The term *WYTIWYG*—what you test is what you get—was coined by Burkhardt et al.[5] to describe the effects of public examination systems on the curriculum. Burkhardt[6] claims that desirable changes in the mathematics curriculum can be brought about through modest, carefully planned changes in examinations. In this way, WYTIWYG can serve as a lever of educational reform.

Some contend that the power of that lever depends on the importance that has been placed on the test results. Popham[7] used the expression *measurement-driven instruction* to describe classroom practices motivated by consequences, or stakes, attached to the test results. He identified two types of high stakes for tests. One type is characterized by the use of scores to make important decisions about students, such as promotion to the next grade, reward of course credit, or qualification for a high school diploma. The other type of high stakes is associated with news media reports of school or district test results. Thus, high-stakes tests draw their power from educators' concerns for students' welfare and for their own standing in the community. One of Madaus's principles of measurement-driven instruction[8] is that high-stakes tests have the power to transfer what was once local control over the school curriculum to the agency responsible for the examination.

We begin, therefore, with an examination of evidence that externally mandated tests are influencing school mathematics curricula and, if so, what the nature of that influence is.

CHANGES IN CONTENT

Several recent studies have looked at the effects that mandated testing programs have on curriculum content. They show that, to various degrees, the WYTIWYG phenomenon is at work in classrooms. According to Stake and Theobold,[9] the most frequently reported change in school conditions that teachers attributed to the increased emphasis on testing is greater pressure to teach stated goals. Darling-Hammond and Wise[10] collected data from in-depth interviews with 43 randomly selected teachers from three large school districts in three mid-Atlantic states. When asked what impact standardized tests had upon their classroom behavior, the most common response was that they changed their curriculum emphasis. Some teachers reported that the emphasis on standardized tests has caused them "to teach skills as they are tested instead of as they are used in the real world."[11]

Nature of the effects   As part of a study of the impact of the minimum competency testing program that the state of Kansas implemented in 1980 at grades 2, 4, 6, 8, and 10, Glasnapp et al.[12] solicited the opinions of school board members, superintendents, principals, and teachers. A different questionnaire was used for each group, with 1,358 teachers participating in the 1982 survey, 816 in 1983, and 1,244 in 1987. The data show that as the test objectives were more widely distributed and as the teachers reported increased encouragement to direct their instruction to the state objectives, there was a corresponding increase in teachers' reports that the test was influencing their instruction. Over half the teachers surveyed in 1987 reported that the test objectives were valuable for identifying what needed to be taught and that they had given those objectives increased emphasis in their instruction. Nearly half said they used the state-distributed minimum competency objectives to plan instructional activities, up from 38 percent in 1983 and 23 percent in 1982. The Kansas teachers also reported that the state minimum competency testing reduced the time they spent teaching skills that the tests did not cover.

Smith and Rottenberg[13] interviewed 19 elementary school teachers and then observed the classes of four of these teachers for an entire semester, during which externally mandated tests were administered. The researchers noted a definite trend, which they attributed to time constraints and a packed curriculum, to neglect topics not included on the standardized tests and to focus on those that were. Mathematics beyond what was to be covered on the tests was given very little attention.

Romberg et al.[14] undertook a study of eighth-grade teachers' perceptions of the impact that their state- or district-mandated testing programs had on mathematics instruction. A national sample of 552 teachers responded to the survey questionnaire. Of the 252 respondents who said they administered a state-mandated test, 34 percent reported placing a greater emphasis on the topics emphasized on the test and 16 percent reported placing less emphasis on topics not emphasized. In reaction to their state testing programs, 23 percent of the teachers were placing a greater emphasis on paper-and-pencil computation whereas only 1 percent were decreasing that emphasis. As for familiarity with their state tests, 46 percent said they looked at the test to see whether the topics were those they were teaching whereas 33 percent reported that they did not examine the test at all.

Strength of the effects Clearly, these studies demonstrate that mandated testing is having an impact on the content of mathematics instruction, but the strength of that impact is another question. Porter et al.,[15] in a review of several studies of elementary school teachers' decisions about instructional content, found little evidence that standardized tests given once a year significantly influence the choices made about what to teach.

These studies did not, however, take into account the importance the elementary school teachers attached to the tests, either for their students or for themselves. Recent studies seem to bear out the claim of Madaus[16] and Popham[17] that the higher the stakes attached to the test results, the greater the impact of the testing program on the curriculum. In interviews conducted by Darling-Hammond and Wise,[18] teachers typically reported that when tests are used to measure teacher effectiveness or student competence, incentives are created to teach the precise test content instead of underlying concepts or untested content.

Corbett and Wilson[19] studied the effects of state-mandated minimum competency testing programs in Maryland and Pennsylvania. At the time of the study, Maryland students needed passing scores on the reading and mathematics tests in order to receive a high school diploma. In Pennsylvania the purpose of the minimum competency tests in language and mathematics was to identify students in need of remedial instruction. Thus the Maryland test was considered by the researchers to have higher stakes than the Pennsylvania test. The study results consistently showed that the Maryland testing program had the more powerful influence on the school curriculum. For example, in Maryland 53 percent of the educators surveyed reported a major or total change in class content resulting from their state testing program. In Pennsylvania only 7 percent reported a major change in their instructional content.

A national survey of teachers on the influence of mandated testing on mathematics and science teaching was conducted as part of a larger study by the Center for the Study of Testing, Evaluation, and Educational Policy.[20] The survey findings were based on over 1800 responses from teachers whose classes were given mandated tests in mathematics. The results showed that teachers with high-minority classes (greater than 60 percent minority students) perceived standardized tests to be of greater importance than did teachers with low minority classes (less than 10 percent minority students). Teachers of high-minority mathematics classes were more likely to use mandated test scores to place students in special services, to recommend students for graduation, and to evaluate student progress. These teachers also felt more pressure to improve their students' scores on mandated mathematics tests. Two thirds of the high-minority classroom teachers said their students' scores on mandated mathematics tests were below their districts' expectations, compared with only one fifth of the teachers of low minority classes. Three quarters of the teachers of high-minority classes agreed that they felt pressure from their districts to improve their students' scores on mandated mathematics tests. Asked about the influences that mandated standardized tests have on their instructional practice, teachers of high-minority classes indicated stronger curriculum effects than did teachers of low-minority classes. Teachers of high-minority classes were more likely to be influenced by mandated tests in their choice of topics and in the emphasis they gave those topics in their mathematics classes.

Direction of the effects Although some researchers have tried to determine whether mandated testing is causing a shift in curriculum content, others have tried to discern the direction of the shift. Shepard and Smith[21] reported, from interviews with and observations of kindergarten and first-grade teachers, that standardized tests at third and sixth grades have served to fix requirements for the end of the first grade. In a position paper on appropriate guidelines for curriculum content and assessment programs, the National Association for the Education of Young Children and the National Association of Early Childhood Specialists in State Departments of Education[22] point to the overemphasis on test results as causing a downward shift in content, so that what used to be taught in first grade is now taught in kindergarten. The impact of such testing has even trickled down into programs for 3- and 4-year-old children.

Many of the state testing directors interviewed by Shepard[23] emphasized that it is the conscious purpose of state testing programs to ensure that essential skills are taught. Several recent studies indicate that the state programs are achieving some success in steering instruction towards basic skills.

Stake and Theobold[24] surveyed 285 teachers in Illinois, Indiana, Minnesota, South Dakota, Washington, Maryland, and North Carolina. When asked for a summary judgment on formal standardized testing, 36 teachers indicated that testing is helpful in many ways and 173 said it is a generally positive factor that is more helpful than harmful. Asked for the single most positive contribution that testing makes in their school, they most often cited the increased time spent teaching basic skills. Corbett and Wilson[25] found that 85 percent of the Maryland educators and 30 percent of the Pennsylvania educators surveyed perceived at least a moderate spread of basic skills instruction throughout the curriculum as a result of their state minimum competency testing programs. Nearly 80 percent of the teachers surveyed by Lomax et al.[26] either agreed or strongly agreed that mandated testing influences teachers so that they spend more instructional time in mathematics classes on basic skills.

Some have concluded from these studies that mandated standardized tests are causing school curricula to move towards an emphasis on basic skills. Archbald and Porter,[27] however, are not so sure. They contend that mandated testing, rather than causing instruction to focus on basic skills, is merely consistent with the instructional practice that would take place in any case. Their skepticism is supported by research findings indicating that teachers have a positive view of teaching basic skills. Research by Glasnapp et al.[28] found that 89 percent of the Kansas teachers surveyed were satisfied or extremely satisfied with their district's emphasis on basic skills instruction. Even though 86 percent of the teachers who participated in the study by Romberg et al.[29] characterized state tests as primarily tests of basic or essential skills, only 31 percent said they placed a greater emphasis on basic skills than they would otherwise. Whether mandated testing programs are the cause or merely a contributing factor, the important point is that the resulting emphasis on basic skills is certainly far from the mathematics curriculum called for by the National Council of Teachers of Mathematics (NCTM) in their *Curriculum and Evaluation Standards*.[30]

There is also evidence that an emphasis on problem solving and critical thinking, which is in line with the NCTM *Standards*, is on the rise as well. Most of the teachers surveyed by Stake and Theobold,[31] 215 of 285, report that an increase in emphasis on problem solving and critical thinking has taken place in their schools over the last year or two. When asked what changes they thought were at least partly caused by an emphasis on testing, one of the three changes most frequently noted was a gain in emphasis on problem solving and critical thinking.

In the study by Romberg et al.,[32] 81 percent of the teachers reported that they knew problem-solving items were on the state test. Whereas 20 percent reported placing a greater emphasis on problem solving because of the state test, only 8 percent reported less emphasis. The researchers suspected, however, that the "teachers who consider problem solving to be on the test are probably thinking of simple word problems"[33] and do not hold the broader conception of problem solving called for in the NCTM *Standards*.

CURRICULUM ALIGNMENT

Leinhardt and Seewald[34] referred to the extent to which instructional content matches test content as *overlap*. They pointed out that teachers are well aware of the notion that the greater the test overlap in their instructional emphasis, the higher their students' test scores are likely to be. The result, according to Resnick and Resnick,[35] is that "school districts and teachers try to maximize overlap . . . by choosing tests that match their curriculum. When they cannot control the tests, . . . they strive for overlap by trying to match the curriculum to the tests, that is, by 'curriculum alignment'." [36]

Though no recent studies directly address the extent to which teachers recognize and strive for overlap, various research methods have been used to measure overlap indirectly. For example, Freeman et al.[37] conducted year-long case studies of several fourth-grade teachers to analyze their styles of textbook use and to determine how the different styles affected content overlap between the mathematics textbook used and five standardized tests of fourth-grade mathematics. The researchers defined five models of textbook use on the basis of their classroom observations. In every case, a substantial proportion of the problems presented during the teachers' lessons dealt with tested topics.

As noted before, the majority of the Kansas teachers who participated in a 1987 survey[38] found the state's minimum competency test objectives to be a valuable guide for their curriculum and reported changing their instructional emphasis accordingly. As teachers strive to maximize overlap, some observers have expressed concern that the curriculum will eventually narrow until instruction and learning are focused exclusively on what is tested.[39] However, there is little in the way of research to support the claim that the curriculum is actually narrowing in response to mandated testing programs. Only 16 percent of those same Kansas teachers had seen indications that the school curriculum was being narrowed as a result of the state minimum competency tests.[40] In fact, according to Stake and Theobold,[41] 199 out of 285 teachers surveyed reported that a general broadening of the curriculum had taken place in their schools over the last few years.

Perhaps these differences in perspective can be attributed to the different ways in which researchers and teachers interpret "narrowing". To researchers, narrowing refers to teaching to the test. Some teachers, however, appear to interpret "narrowing" as teaching fewer topics. That only a small fraction of Kansas teachers believed that their curriculum had narrowed can perhaps be explained by two other statistics: 45 percent of them reported adding lessons or units to the curriculum as a result of the tests, whereas only 17 percent reported sacrificing instruction in other areas or skills to teach to the state objectives. In response to pressures to alter content, teachers have demonstrated a greater willingness to add topics to the curriculum than to delete them.[42]

The Corbett and Wilson research[43] suggests that a narrower curriculum may be welcomed by some teachers. Even though 64 percent of the Maryland educators surveyed said that there had been at least a moderate narrowing of their school curriculum as a result of the state minimum competency test, 56 percent also reported at least a moderate improvement in the curriculum. In follow-up interviews, Maryland educators said the curriculum was "structured, coordinated, more focused, more defined, sequentially ordered, more systematic, consistent, and created a consciousness (about what was being taught)."[44] For these Maryland teachers, narrowing was associated with bringing an unwieldy curriculum under control.

Mandated testing is also impinging on the teaching methods used in mathematics classrooms. The study of eighth-grade teachers by Romberg et al.[45] offers some insight into the magnitude and nature of the impact. With respect to teaching methods, 18 percent of the teachers reported an increased use of direct instruction and only 1 percent a decreased use as a result of their state mathematics test. Alternatives to direct instruction did not prosper. Although 11 percent increased their use of small group instruction, another 9 percent decreased their small group instruction. Despite increased advocacy of cooperative learning, the 6 percent who reported an increase in its use were offset by the 10 percent reporting a decrease. Similarly, Lomax[46] reported that mathematics teachers believed mandated testing had resulted in more time spent in whole group instruction.

Extended projects have been advocated as a way of engaging students in a deeper and more sustained encounter with mathematics. Mandated testing appears to work against such projects, perhaps because they are seen as taking class time that might be used to prepare for the tests. Only 2 percent of the teachers in the Romberg et al.[47] sample reported an increase in extended project work, whereas 22 percent reported a decreased emphasis on extended projects. A similar phenomenon affects the use of technology in teaching when such technology is not incorporated into mandated testing programs. Whereas 5 percent of the eighth-grade teachers reported increasing their emphasis on calculator activities, 20 percent reported a decreased emphasis.[48] Only 2 percent reported an increased emphasis on computer activities whereas 16 percent reported a decrease in computer activities.

These results indicate that the magnitude of the impact of mandated testing on teachers' instructional methods is rather limited, a finding that is supported by the Glasnapp et al. study.[49] Only 19 percent of the Kansas teachers surveyed in 1987 said they changed their instructional practices because of the state testing program, even though most of them said they used the test results to assess their teaching effectiveness. It is quite possible that the test results gave the Kansas teachers no reason to change their teaching because students scored well on the 1987 mathematics tests. Most school principals perceived student performance on the tests to be at or above expectations. Also, how strongly a mandated testing program influences instructional methods may, as with instructional content, be a function of the test stakes. The Kansas minimum competency test was a low-stakes test.[50] When Corbett and Wilson[51] asked whether teachers had adopted new instructional approaches as a result of the state minimum competency testing program, they found dramatically different responses in high-stakes Maryland compared with low-stakes New Jersey. Nearly twice as many Maryland educators (82 percent) as New Jersey educators reported teachers' methods had changed at least moderately.

Although the strength of the impact of the state tests on instructional practice, as reflected in the opinions of samples of eighth-grade teachers and of Kansas teachers, may not be cause for concern, the direction in which practice is moving is. A greater reliance on direct instruction, accompanied by a de-emphasis on projects, calculators, and computers, is directly opposed to the practices envisioned by the 1991 NCTM *Professional Standards for Teaching Mathematics*.[52]

Some of the reported effects of testing on teaching practice can be considered positive. According to Glasnapp et al.,[53] about half of the Kansas teachers believed that the state-mandated test of minimum competency allowed them to match their instructional methods to the performance levels of individual students. Two thirds believed that the minimum competency tests informed their instructional decisions by increasing their understanding of a student. Over two thirds of the teachers surveyed by Stake and Theobold[54] perceived an increase in the attention given to differences in individual students over the last few years in their school, with many of them attributing the increase at least partly to the emphasis on testing.

TEST PREPARATION

Teachers' test preparation practices give rise to two concerns: the amount of time from regular instruction given to test preparation and the educational value of that preparation. Certainly, some attention to teaching students how to manage testing time and answer sheets is appropriate and can lead to more valid results. As Shepard[55] pointed out, however, repeated practice aimed strictly at the content of the test rather than the content domain of the test can increase scores without increasing student achievement. Teachers' test preparation practices have received much scrutiny and criticism.[56] According to Madaus,[57] teachers respond to the pressures of high-stakes tests by preparing students to meet the requirements of previous test questions, which are reduced to the level of strategies in which the students are drilled.

Three recent research studies have asked teachers about the extent of their test preparation practices. Smith and Rottenberg[58] report that in the four elementary classes they observed, an average of 54 hours of class time was spent preparing for externally mandated standardized tests, which, in addition, required about 18 hours to administer. The survey questionnaire used by Romberg et al.[59] asked the grade 8 teachers to indicate the preparation practices for which they set aside several days a year, several weeks a year, or time on a frequent and regular basis. Most typically, the teachers indicated that they prepared for the state tests only several days a year. In addition, 30 percent of the teachers reported allocating no instructional time for state test preparation and 46 percent reported no time for district test preparation.

The extent of test preparation is undoubtedly affected by several factors. Lomax et al.[60] found the percentage of teachers who spent more than 20 hours of class time on mandated test preparation was three times higher for high-minority classes than for low-minority classes. Also, nearly three quarters of the teachers of high-minority classes began their mathematics test preparations a month or more before the test, well over twice the percentage of teachers of low-minority classes who did so.

Information is also available about the nature of teachers' test preparation practices. In the 1987 survey by Glasnapp et al.,[61] 40 percent of the teachers indicated that the Kansas minimum competency test had led to drills, coaching, and test item practice. Lomax et al.[62] report that the most common practices teachers reported using in preparing students for mandated mathematics tests were teaching test-taking skills (73 percent), encouraging students to work hard (64 percent), teaching topics known to be on the test (50 percent), providing students with items similar to test items (47 percent), and using motivating materials (45 percent).

Are test preparation practices worth the valuable class time they take? Popham[63] asserts that two standards must be met for test preparation activities to be considered suitable for classroom instruction: The practice must meet professional standards of ethics, and it should be of educational value to students. He considers any action that violates test security procedures to be outside the boundaries of professional ethics. Instruction aimed at increasing students' test scores without increasing their mastery of the domain of concepts or skills to be measured by the test is of no educational value.

To gauge the beliefs of educators about test preparation practices, Popham conducted a brief survey of the participants in three workshops he held in late 1989 and early 1990. The first workshop was attended by teachers and administrators from Ohio, Indiana, and Kentucky. Teachers, administrators, and school board members from Southern California took part in the last two workshops. The participants were asked to supply anonymous judgments as to whether five test preparation practices were appropriate or inappropriate: previous-form preparation, current-form preparation, generalized instruction on test-taking skills, same-format preparation, and varied format preparation. The survey included a description of the five practices and of professional ethics and educational defensibility standards. Over a quarter of the teachers said they provided test-specific materials and used practice tests to prepare their students for mandated mathematics tests.

Popham[64] found substantial numbers of teachers in both of his samples who considered previous-form or current-form preparation to be appropriate. More than half of each sample deemed same-format use appropriate. Popham, on the other hand, considers all three practices to be educationally indefensible, and current-form use to be unethical. Although Popham's samples are small, his results suggest that many teachers may be willing to engage in questionable practices to improve their students' test performance.

In a study by Hall and Kleine,[65] 1,012 superintendents, testing coordinators, principals, and teachers responded to a questionnaire on the use of test preparation materials. These educators were from districts across the country where standardized, group-administered, norm-referenced tests were given. Of the 176 teachers surveyed, 55 percent reported using some type of test preparation materials, the most common of which were locally developed.

Mehrens and Kaminski[66] examined four commercially available test preparation programs to see how well they fit the content and format of the California Achievement Test (CAT). They found substantial variation. In the most extreme case, the Test-Specific Scoring High materials were matched so closely to the CAT that it "serves as a pre-test for the CAT in the same manner as if one actually used the CAT as a pretest prior to giving the same CAT at a later time."[67] In the Hall and Kleine study,[68] 19 percent of the teachers using test preparation materials were using the Scoring High materials.

In an ideal world, all of the instructional materials teachers use to prepare students for standardized tests would be interesting tasks of sound instructional value. Research suggests, however, that there is a move toward test preparation practices that are debatable at best, from the standpoint of both professional ethics and mathematics education.

CLASSROOM ASSESSMENT PRACTICES

The question of whether externally mandated testing programs affect a teacher's own assessment practices has been considered by a few researchers. The answer depends on which assessment practice is being considered. Interviews conducted by Salmon-Cox[69] with 68 elementary school teachers revealed that these teachers did not give standardized test scores much attention when assessing their students' progress. When discussing general assessment techniques, the teachers most frequently mentioned observation as well as teacher-made tests and interaction with students. Only three of the teachers spoke of standardized tests when discussing how they assess students.

Other studies, however, show that standardized tests do influence the ways in which teachers design their own tests. Madaus[70] points out that teachers pay particular attention to the form of the questions on a high-stakes test. According to Romberg et al.,[71] more than 70 percent of the grade 8 teachers surveyed perceive that typical district- and state-mandated test items require a single correct answer and are in a multiple-choice format. There is concern in the mathematics education community that such a format offers students no opportunity to gather, organize, or interpret data, to model, or to communicate, all of which are called for in the

NCTM *Standards*.[72] The form of the questions widely used on standardized tests implies to students that their task is not to engage in interpretive activity, but to find, or guess, in a quick, nonreflective way, the single correct answer that has already been determined by others.[73] If you do not know an answer immediately, there is no way of arriving at a sensible response by thought and elaboration. In the study by Darling-Hammond and Wise,[74] teachers reported feeling the need to use similar types of test items and fewer essay tests in their own assessment practice.

Only 13 percent of the teachers in the study by Romberg et al.[75] reported that they were not familiar with the format and style of typical test items on the state test. Over a third considered the format and style of state-mandated test items when planning their instruction. On a short one-page questionnaire sent to the teachers who did not send back the first survey, 51 percent of the 142 respondents reported considering the style and format of test items when planning their own tests.

Teachers' assessment tools provide a well-defined medium for indicating to students what it is about mathematics that is most important. Because of the importance of teachers' tests and the indications from research that teachers see standardized tests as a model of what those tests should be, further study of teachers' testing practices is critical.

- -

EFFECTS ON TEACHERS

Research studies give conflicting reports of how teachers feel about and react to mandated testing programs. According to Glasnapp et al.,[76] 62 percent of the Kansas teachers surveyed in 1987 either agreed or strongly agreed that the pressure on local districts to perform well on the state minimum competency tests led to undesirable educational practices. In the same survey, however, 60 percent of the teachers indicated that they considered the Kansas minimum competency testing, overall, to be beneficial to education in the state. The survey by Lomax et al.[77] also gives conflicting views from teachers. Although over half of the teachers agreed that the mandated testing program in mathematics led to teaching methods that went against what they considered good instructional practice, a substantial portion, 28 percent, agreed that mandated testing helps schools achieve the goals of the current education reform movement. Most of the teachers surveyed by Stake and Theobold[78] judged standardized testing to be a generally positive influence on the quality of education at their schools.

Teachers are apparently able to separate what they may see as generally favorable influences of testing on education from how they are responding in their own classes to mandated testing. The teachers in the Darling-Hammond and Wise study[79] commented that they felt that teaching to standardized tests is not really teaching. Teachers committed to developing a new school curriculum at an elementary school observed by Livingston et al.[80] and at a magnet high school observed by McNeil[81] had a common reaction to the demands of mandated testing. Livingston et al.[82] reported the experiences of Westwood School, a K–2 school in Dalton, Georgia, where the teachers undertook a revision of the state mathematics curriculum. As the changes proceeded, the teachers became concerned that the content and format of the state-mandated standardized tests might not reflect their students' experiences with the revised curriculum. The teachers believed that their desire for innovation was constrained by the need to teach to the test. The perceived conflict between the new curriculum and the curriculum based on mandated test objectives was resolved at Westwood by attempting to teach both curricula simultaneously.

McNeil[83] observed the experiences of teachers at an innovative magnet high school as the district piloted a system of proficiency examinations to be administered at the end of each semester of coursework. Test results were linked to teacher merit pay and principals' bonuses. School scores were compared in the newspapers. The teachers responded by finding ways of working around the proficiencies, believing that they were too confining. The teachers coped with the pressure to teach the objectives of the proficiency-based curriculum by delivering what McNeil calls "double-entry" lessons, in which the lessons geared to the proficiency lessons were delivered in addition to the regular course instruction.[84]

Smith and Rottenberg[85] observed some negative effects of mandated testing on teachers. The elementary school teachers in their study expressed feelings of shame and embarrassment if their students scores were low or did not meet district standards. The researchers noted a sense of alienation resulting from teacher beliefs that test scores are more a function of students' socioeconomic status and effort than of classroom instruction. Also, the teachers believed that test results were not properly interpreted in the community, where low scores were attributed to weak school programs and lazy teachers.

The Westwood School curriculum committee members contended that discrepancies between teachers' judgments and students' test scores lead to a deprofessionalization of teachers because of the view by parents that test scores are absolute indicators of students' learning.[86] Of the Maryland teachers surveyed by Corbett and Wilson,[87] 58 percent reported that mandated testing has led to at least a moderate decrease in professional judgment in instructional matters. Included in the survey instrument were questions on the effects of mandated testing on teachers' work life: 70 percent of the respondents reported a major increase in demands on their time, 66 percent a major increase in paperwork, 64 percent a major increase in pressure for student performance, 55 percent at least moderate changes in staff reassignment, and 44 percent at least a moderate increase in worry about lawsuits.

These studies do not present a positive picture of the impact of mandated testing on the teachers who administer them. Tests required by agencies outside of the classroom have added to teachers' work loads. Meanwhile, tying students' scores to promotion or course credit decisions has taken away from teachers' authority. As the research of McNeil[88] and of Livingston et al.[89] shows, mandated standardized testing programs can pose a real dilemma for teachers who want to implement changes in the curriculum. As long as the mathematics content of standardized tests differs from the mathematics curriculum called for in the NCTM *Standards,* teachers faced with mandated testing will find themselves in a difficult position. For example, even teachers who recognize the benefits of calculators often justify their reluctance to use them in their mathematics classes by arguing that students are not allowed to have them while taking standardized tests.[90]

· ·

EFFECTS ON STUDENTS

There is evidence that standardized test scores play a major role in determining students' educational experiences. According to Salmon-Cox,[91] about one quarter of the elementary school teachers she interviewed reported that they consider standardized test results, in conjunction with other information, in grouping and tracking students. Smith and Rottenberg[92] also observed that test scores were used to make decisions about placement of students into groups and tracks. They note that test scores were the single most important factor used in the decision to place children into gifted programs and into an advanced junior high school curriculum. In their study, Romberg et al.[93] report that 35 percent of the eighth-grade teachers surveyed indicate that district-mandated tests influence decisions about grouping students within the class for instruction, and 62 percent say the district test scores influence recommendations of students for course or program assignments.

Mandated tests can have a negative impact on the students who take them. The National Association for the Education of Young Children and the National Association of Early Childhood Specialists in State Departments of Education[94] hold that many young children experience unnecessary frustration as they struggle with developmentally inappropriate standardized tests for kindergarten and first grade. Smith and Rottenberg[95] note that most teachers believe the frequency and nature of the tests and the way in which they are administered cause "stress, frustration, burnout, fatigue, physical illness, misbehavior and fighting, and psychological distress,"[96] particularly in younger students.

On the other hand, Corbett and Wilson[97] note some positive effects of mandated testing on students. Their survey asked teachers about their perceptions of the impact of the state minimum competency tests on students' work life. When asked if students had become more serious about their classes, 40 percent of the Maryland teachers indicated that they perceived at least a moderate change in this direction. Also, most of the Maryland teachers reported at least a moderate increase in their empathy for students who are poor achievers and in their knowledge of students with serious learning problems. The Maryland teachers were able to turn what they reported as a mostly negative influence on their own work lives into what they saw as a more positive experience for their students.

CONCLUSIONS

The movement for accountability in education that has increased the volume of mandated testing in American schools in the past two decades has come more from a desire to find out what students are learning than from a demand to change the content of that learning or to shape its acquisition. Nonetheless, raising the level of academic performance has been part of the agenda from the outset. Rewards and punishments are built into each system of mandated testing, whether what is at stake is graduation from high school or a headline in the local paper.

Because most of the mandated mathematics testing has concentrated on basic concepts and manipulative skills that can be assessed by multiple-choice or short-answer tests, any effect of that testing on the school curriculum has been to increase the already substantial attention teachers give to such concepts and skills. Teachers have not necessarily found such a narrowing of the curriculum to be bad: It allows them to direct their attention to topics some authority considers important, and it is in line with what they are likely to feel comfortable teaching. Their instructional practice has, if anything, shifted to an even greater reliance on direct instruction, which is marked by organized lessons presented through lecture and discussion to the entire class. Because teachers are familiar with direct instruction and usually feel comfortable using it, they may see the effects of mandated testing on their teaching as positive.

Teachers are not necessarily comfortable, however, with everything that mandated testing may require of them. They may feel called upon to use valuable class time preparing their students for the tests, at times engaging in practices with dubious educational or ethical value. They may feel that testing programs devalue their skills as assessors of students' learning. When the demands of mandated testing programs conflict with practices they deem more appropriate, however, they tend not to challenge these programs. Rather, they seek a middle ground in which they strive to meet both the demands of the testing program and their own view of what and how they should be teaching. Even as reports surface of some undesirable effects on students and their educational opportunities, teachers continue to see as many benefits as flaws in mandated testing. They have accommodated to the system.

The available research does not lead to the unqualified conclusion that mandated testing is having harmful effects on mathematics instruction. The picture is both more mixed and more indistinct. One reason may be that the research has focused on teachers' views of mandated testing and has seldom pushed very far beyond what teachers have said on questionnaires or in response to interview questions. Teachers may not be all that dissatisfied with testing programs that are largely in tune with their styles and beliefs. Moreover, they may not see effects that could be detected through other means.

A hint that there may be more to the story comes from the study by Stake and Theobold.[98] Although none of the teachers in their sample said that the description of their school, as presented in the survey answers, was in any way biased or misleading, Stake and Theobold drew the following conclusion: "We are not satisfied with the data presented here. We do not believe these data tell us what is happening to schooling in America."[99] Stake and Theobold contend that teachers—like the rest of us—lack a language for representing the curriculum so as to distinguish personal concepts of education from the official indices provided by learning objectives and test items. In other words, teachers may not be able to tell us clearly about the effects of mandated testing. As we work to develop a richer language to describe the curriculum, we need also to consider means of investigating the effects of mandated testing that do not rely exclusively on teachers' reports.

The WYTIWYG phenomenon is clearly quite limited as an explanation of how mandated testing produces effects on mathematics instruction and, therefore, learning. Students always learn some mathematics that is not tested, and they do not always learn all the mathematics being tested. In addition to portraying the student as a faultless receptacle for instruction, WYTIWYG leaves out the teacher as a medium for turning test prescriptions into learning experiences. As Silver[100] observed, "perhaps WYTIWYG should be more accurately dubbed WYGIWICT—what you get is what I can teach."[101]

Silver's point becomes especially important as some mandated testing programs change to incorporate reforms sought in such documents as the NCTM *Standards*. As these programs incorporate items with extended answers, calling upon students to perform investigations of open-ended problems, write up their findings, and perhaps collect them in a portfolio, the intention is that teachers will be drawn away from a basic skills curriculum in mathematics that is delivered through direct instruction. Teachers will almost certainly find open-ended work more difficult to manage than direct instruction and an ambitious curriculum more difficult to implement than basic skills. The limitations on their ability and their willingness to teach in the ways sought by reformers will then begin to govern how the mandated testing affects their instruction. We may begin to see some teachers challenging or attempting to subvert a system of assessment that suits neither their teaching style nor their beliefs about essential mathematics content.

The effects of mandated testing on instruction have not been well studied and are not clear. Furthermore, changes currently under way in mandated testing may modify whatever effects there are. The picture given by the available research is neither so bleak as the one advanced by foes of standardized, multiple-choice testing nor so rosy as that offered by proponents of testing as the engine of school reform. It is instead a blurred picture. Improvements in research techniques and more extensive investigations may ultimately yield a more focused view. The landscape of mathematics instruction and assessment is itself changing, and even the tentative conclusions we have drawn in this paper seem unlikely to hold for long.

ENDNOTES

[1] U.S. Congress, Office of Technology Assessment, *Testing in American schools: Asking the right questions* (OTA-SET-519) (Washington, DC: U.S. Government Printing Office, 1992), 3-4.

[2] R. E. Stake and P. Theobold, "Teachers' views of testing's impact on classrooms," in *Advances in program evaluation: Effects of mandated assessment on teaching*, ed. R. E. Stake and R. G. O'Sullivan, (Vol. 1, Part B) (Greenwich, CT: JAI Press, 1991), 189-201.

[3] Lauren B. Resnick and Daniel P. Resnick, "Assessing the thinking curriculum: New tools for educational reform," in *Changing assessments: Alternative views of aptitude, achievement, and instruction*, ed. B. R. Gifford and M. C. O'Connor, (Washington, DC: National Commission on Testing and Public Policy, 1991), 37-75.

[4] Ibid., 59.

[5] Hugh Burkhardt, R. Fraser, and J. Ridgway, "The dynamics of curriculum change, in *Developments in school mathematics education around the world* (Proceedings of the Second UCSMP International Conference on Mathematics Education, 7-10 April 1988), ed. I. Wirszup and R. Streit, (Reston, VA: National Council of Teachers of Mathematics, 1990), 3-30.

[6] Hugh Burkhardt, "Curricula for active mathematics," in *Developments in school mathematics education around the world* (Proceedings of the UCSMP International Conference on Mathematics Education, 28-30 March 1985), ed. I. Wirszup and R. Streit, (Reston, VA: National Council of Teachers of Mathematics, 1987), 321-361.

[7] W. J. Popham, "The merits of measurement-driven instruction," *Phi Delta Kappan, 68,* (1987), 679-682.

[8] George F. Madaus, "The influence of testing on the curriculum," in *Critical issues in curriculum* (87th Yearbook of the National Society for the Study of Education, Part 1), ed. L. N. Tanner, (Chicago: University of Chicago Press, 1988), 83-121.

[9] "Teachers' Views."

[10] Linda Darling-Hammond and Arthur E. Wise, "Beyond standardization: State standards and school improvement," *Elementary School Journal, 85,* (1985), 315-336.

[11] Ibid., 320.

[12] D. R. Glasnapp, J. P. Poggio, and M. D. Miller, "Impact of a 'low stakes' state minimum competency testing program on policy, attitudes, and achievement," in *Advances in program evaluation: Effects of mandated assessment on teaching*, ed. R. E. Stake and R. G. O'Sullivan (Vol. 1, Part B, pp. 101-140), (Greenwich, CT: JAI Press, 1991).

[13] M. L. Smith and C. Rottenberg, "Unintended consequences of external testing in elementary schools," *Educational Measurement: Issues and Practice, 10*(4), (1991), 7-11.

[14] Thomas A. Romberg, E. A. Zarinnia, and S. R. Williams, *The influence of mandated testing on mathematics instruction: Grade 8 teachers' perceptions,*

(Madison: University of Wisconsin-Madison, National Center for Research in Mathematical Science Education, 1989).

[15] Andrew C. Porter, R. Floden, D. Freeman, W. Schmidt, and J. Schwille, "Content determinants in elementary school mathematics," in *Perspectives on research on effective mathematics teaching*, ed. Douglas A. Grouws and Thomas J. Cooney, (Hillsdale, NJ: Lawrence Erlbaum, 1988), 96-113.

[16] "Influence of Testing."

[17] "Measurement-driven instruction."

[18] "Beyond Standardization."

[19] H. D. Corbett and B. L. Wilson, *Testing, reform, and rebellion*, (Norwood, NJ: Ablex, 1991).

[20] R. G. Lomax, *The influence of testing on teaching math and science in grades 4-12, Appendix A: Nationwide teacher survey*, (Chestnut Hill, MA: Boston College, Center for the Study of Testing, Evaluation, and Educational Policy, 1992); R. G. Lomax, M. M. West, M. C. Harmon, K. A. Viator, and G. F. Madaus, *The impact of mandated testing on minority students*, (Chestnut Hill, MA: Boston College, Center for the Study of Testing, Evaluation, and Educational Policy, 1992).

[21] L. A. Shepard and M. L. Smith, "Escalating academic demand in kindergarten: Counterproductive policies," *Elementary School Journal, 89,* (1988) 135-145.

[22] National Association for the Education of Young Children and the National Association of Early Childhood Specialists in State Departments of Education, "Guidelines for appropriate curriculum content and assessment in programs serving children ages 3 through 8: Position statement," *Young Children, 46(3),* (1991), 21-38.

[23] L. A. Shepard, "Inflated test score gains: Is it old norms or teaching the test?", (Paper presented at the annual meeting of the American Educational Research Association, San Francisco, 1989). (ERIC Document Reproduction Service No. ED 334 204).

[24] "Teachers' Views."

[25] *Testing, reform, and rebellion.*

[26] *Impact of mandated testing.*

[27] D. A. Archbald and A. C. Porter, *A retrospective and an analysis of roles of mandated testing in education reform,* paper prepared for the Congressional Office of Technology Assessment (Washington, DC, 1990).

[28] "Impact of 'low stakes' testing."

[29] *Mandated testing.*

[30] National Council of Teachers of Mathematics, *Curriculum and evaluation standards for school mathematics*, (Reston, VA: Author, 1989).

[31] "Teachers' views."

[32] *Mandated testing.*

[33] Ibid., 84.

[34] G. Leinhardt and A. M. Seewald, "Overlap: What's tested, what's taught?" *Journal of Educational Measurement, 18,* (1981), 85-96.

[35] "Thinking curriculum."

[36] Ibid., 57.

[37] D. J. Freeman, G. M. Belli, A. C. Porter, R. E. Floden, W. H. Schmidt, and J. R. Schwille, "The influence of different styles of textbook use on instructional validity of standardized tests," *Journal of Educational Measurement, 20,* (1983), 259-270.

[38] "Impact of `low stakes' testing."

[39] "The influence of testing."

[40] "Impact of `low stakes' testing."

[41] "Teachers' views."

[42] R. E. Floden, A. C. Porter, W. H. Schmidt, D. J. Freeman, and J. B. Schwille, "Responses to curriculum pressures: A policy-capturing study of teacher decisions about content," *Journal of Educational Psychology, 73,* (1981), 129-141.

[43] *Testing, reform, and rebellion.*

[44] Ibid., 71.

[45] *Mandated testing.*

[46] *Influence of testing.*

[47] *Mandated testing.*

[48] Ibid.

[49] "Impact of `low stakes' testing."

[50] Ibid.

[51] *Testing, reform, and rebellion.*

[52] National Council of Teachers of Mathematics *Professional Standards for Teaching Mathematics,* (Reston, VA: Author, 1991).

[53] "Impact of `low stakes' testing."

[54] "Teachers' views."

[55] "Inflated test score gains."

[56] See, e.g., J. J. Cannell, *How public educators cheat on standardized achievement tests,* (Albuquerque, NM: Friends for Education, 1989).

[57] "The influence of testing."

[58] "Unintended consequences."

[59] *Mandated testing.*

[60] *Impact of mandated testing.*

[61] "Impact of `low stakes' testing."

[62] *Impact of mandated testing.*

[63] W. J. Popham, "Appropriateness of teachers' test-preparation practices," *Educational Measurement: Issues and Practice, 10*(4), (1991), 12-15.

[64] Ibid.

[65] J. L. Hall and P. F. Kleine, *Preparing students to take standardized tests: Have we gone too far?* (Oklahoma City: University of Oklahoma, 1990). (ERIC Document Reproduction Service No. ED 334 249).

[66] W. A. Mehrens and J. Kaminski, "Methods for improving standardized test scores: Fruitful, fruitless, or fraudulent?" *Educational Measurement: Issues and Practice, 8*(1), (1989), 14-22.

[67] Ibid., 18.

[68] *Preparing students.*

[69] L. Salmon-Cox, "Teachers and standardized achievement tests: What's really happening?" *Phi Delta Kappan, 62,* (1981), 631-634.

[70] "The influence of testing."

[71] *Mandated testing.*

[72] Mathematical Sciences Education Board and Board on Mathematical Sciences, National Research Council, *Everybody counts: A report to the nation on the future of mathematics education,* (Washington, DC: National Research Council, 1989); Jean Kerr Stenmark, ed., *Mathematics assessment: Myths, models, good questions, and practical suggestions,* (Reston, VA: National Council of Teachers of Mathematics, 1991).

[73] "Thinking curriculum."

[74] "Beyond standardization."

[75] *Mandated testing.*

[76] "Impact of `low stakes' testing."

[77] *Impact of mandated testing.*

[78] "Teachers' views."

[79] "Beyond standardization."

[80] C. Livingston, S. Castle, and J. Nations, "Testing and curriculum reform: One school's experience," *Educational Leadership, 46*(7), (1989), 23-25.

[81] L. M. McNeil, "Contradictions of control: Part 3, Contradictions of reform," *Phi Delta Kappan, 69,* (1988), 478-485.

[82] "Testing and curriculum reform."

[83] "Contradictions of control."

[84] Ibid, 483.

[85] "Unintended consequences."

[86] "Testing and curriculum reform."

[87] *Testing, reform, and rebellion.*

[88] "Contradictions of control."

[89] "Testing and curriculum reform."

[90] J. W. Kenelly, ed., *The use of calculators in the standardized testing of mathematics*, (New York & Washington, DC: College Board & Mathematical Association of America, 1989).

[91] "Teachers and standardized achievement tests."

[92] "Unintended consequences."

[93] *Mandated testing.*

[94] National Association for the Education of Young Children and the National Association of Early Childhood Specialists in State Departments of Education, "Guidelines."

[95] "Unintended consequences."

[96] Ibid., 10.

[97] *Testing, reform, and rebellion.*

[98] "Teachers' views."

[99] Ibid., 200.

[100] Edward A. Silver, "Assessment and mathematics education reform in the United States," *International Journal for Educational Research, 17,* (1992), 489-502.

[101] Ibid., 500.

DESIGN INNOVATIONS IN MEASURING MATHEMATICS ACHIEVEMENT

STEPHEN B. DUNBAR
UNIVERSITY OF IOWA
ELIZABETH A. WITT
UNIVERSITY OF KANSAS

Nearly a century ago, a movement was afoot in American education, a movement with its origins in a prevailing perception among educators and public alike that our schools were failing to provide the leadership needed to prepare the next generation for the twentieth century. In a retrospective commentary on that movement and the uses and abuses of examinations in the pursuit of the educational reform efforts of that movement, McConn[1] described the avowed purpose of nearly all achievement testing at the time as ensuring

> the *maintenance of standards*, including, as already noted, the enforcement of both prescribed subject matter and of some more or less definitely envisaged degree of attainment.

> If one is to raise any objections here, he must tread softly, because he is approaching what is to many educators in service, especially many of the older ones, the Ark of the Covenant. When those of us who are now in our forties and fifties were learning our trade, "Standards" was the great word, the new Gospel, in American education. To set Standards, and enforce Standards, and raise Standards, and raise them ever more, was nearly the whole duty of teachers, principals and presidents."[2]

McConn goes on to discuss the various unanticipated outcomes of the movement toward what he called Uniform Standards of the Nineties, the principal one being a nearly complete lack of concern for individual differences and their importance in the development of differentiated standards. A second was the use of tests as exclusionary devices rather than instruments that could potentially guide individual students and teachers to more effective approaches to instruction and learning at all levels of education.

McConn's remarks set an appropriate context for the present discussion of design innovations in the development and evaluation of large-scale measures of achievement in mathematics; the clear impetus for new approaches is very similar to the concern for standards that was raised at the end of the nineteenth century. Perhaps it is mere coincidence that such issues come into clear focus at the end of a century. That is a debate for historians. What is at issue in this paper is the perspective from which we approach the very real concern that American society has voiced regarding the preparation of students for a way of life and work that relies increasingly on technological innovations and the ability to think critically and solve problems of a technical nature.

Mathematics instruction is presently seen as a principal vehicle through which American schools will prepare students in this domain, so it is clearly appropriate to consider the role of new tests in enhancing mathematics education. It is equally important to recognize the possible contradiction between the ideals of diversified approaches to assessment on the one hand and the specification of uniform standards of achievement for all students on the other. History does tell us that the primacy of the latter can completely undermine the anticipated benefits of the former, and today's rhetoric on standards is characterized by a uniformity of goals of instruction, albeit a well-intentioned uniformity.

Unlike the press for educational standards at the turn of the last century, the current movement for innovation in methods of measuring achievement in mathematics has the benefit of extensive experience in measuring achievement on a large scale. Although accountability remains the focus of many who are interested in using tests to monitor educational reform efforts, the profession is mindful of expanded definitions of the criteria traditionally used to evaluate measures of achievement. Content quality and cognitive complexity, generalizability and transfer, content coverage and meaningfulness, consequences and fairness, cost and efficiency are some of the criteria that have been recently proposed to characterize expectations for traditional and new forms of assessment and validation of their appropriateness[3]. This paper addresses the problems and issues faced in attempting to satisfy these criteria when developing and implementing new mathematics assessment procedures.

The multifaceted approach to test validation advocated by Linn et al.[4] encourages a less mechanistic approach to investigations of validity and forces a confrontation with what might be called the "broad brush syndrome" in educational assessment. In education and policy circles, there is a strong tendency to paint pictures of critical issues with an extremely broad brush. A given test is either valid or invalid. A performance-based approach to measuring achievement in mathematics will improve student learning of strategies for solving complex, real-world problems. Ratings of portfolios are inherently unreliable. Multiple-choice items cannot measure higher-order skills. Statements like these are symptomatic of the broad brush syndrome, a way of thinking that sees all tests of a given type as alike in their inherent attributes and influences on educational process.

To advance the debate about the role of assessment in the improvement of instruction and learning in mathematics is to approach the easel with a full palette and array of tools. This means facing the fundamental questions of validity for all types of assessments and understanding the importance of consequences, intended and unintended, in the overall evaluation of both traditional and innovative approaches to the design of instruments. Some of this effort can be exerted in the process of developing alternative measures of mathematics achievement themselves, but the effort requires that developers establish empirical grounds for the consequential and evidential bases for test use.[5] Exemplary projects carried out on a national scale can provide some empirical evidence in this regard. The extent to which their results can be generalized to yet to be designed systems for large-scale assessment of mathematics achievement is a key consideration, however.

CONTENT CONSIDERATIONS FOR MATHEMATICS ACHIEVEMENT

By and large, the data available from large-scale, performance-based assessments of educational achievement come from operational assessment programs in the area of direct writing assessment.[6] Writing is an area with a long tradition of performance-based measures being used either to supplement or supplant multiple-choice tests of formal English grammar. Although curriculum specialists argue over the extent to which timed writing samples can support general inferences about achievement in writing, writing samples have been generally regarded as critical to comprehensive assessment programs in the language arts. Over time, the content domains sampled in direct writing assessments have become organized around traditional rhetorical modes of discourse, and the content specifications for the development of writing tasks and scoring protocols in many testing programs reflect an evolved conception of domains to be sampled.[7]

In considering anticipated features of innovative assessments in mathematics, the definitions of content domains should be carefully evaluated. Traditionally, mathematics has been regarded by test developers as an area in which substantial consensus existed with regard to content and the sequencing of subject matter. However, with the introduction of the National Council of Teachers of Mathematics (NCTM) *Standards* [8] for mathematics curricula, the domain of the mathematics teacher has been expanded considerably, and what was once a clear scope and sequence subject for teachers is in the process of being redefined and, as some suggest, "conceptualized as an ill-structured discipline." [9]

When the NCTM *Standards* are used as guidelines for the design of innovative approaches to assessment, complications for measurement arise as a result of the interdisciplinary nature of the standards and the media through which certain standards may be amenable to assessment. For example, Standard 2 describes various ways in which mathematics is used to communicate ideas, and the statements of objectives include such phrases as "reflect and clarify thinking," "relate everyday language to mathematical language," and "representing, discussing, reading, writing, and listening to mathematics." Standard 3 (mathematics as reasoning) emphasizes the importance of explaining mathematical thinking and justifying answers. The goals for instruction reflected by these standards entail an integration of formal mathematical thinking with more generalized reasoning and problem solving throughout the school curriculum, generally observed by teachers through verbal interactions with learners. A more integrated approach to curriculum design is seen as critical for the development of higher-order thinking skills that will be required in an increasingly technological society. The implications for measurement have to do with the role of more generalized cognitive skills in the observable outcomes of mathematics learning.

The QUASAR project described by Lane[10] illustrates one formal attempt to use the NCTM *Standards* as the principal basis for structuring a large-scale nontraditional assessment of mathematics achievement. The project was created to demonstrate the feasibility of implementing programs based on the NCTM *Standards* in middle schools located in economically disadvantaged communities. Development efforts in the QUASAR project focused on the specification of a content framework for both tasks and scoring rubrics. The content framework for the performance tasks can be understood in traditional terms as a table of specifications with process and content dimensions; however, the dimensions of the QUASAR blueprint for task development are not isomorphic with those used in traditional test development for achievement in mathematics. In addition to a more detailed explication of the cognitive processes associated with mathematics content, the QUASAR frameworks included dimensions for mode of problem presentation (e.g., written, pictorial, graphic, arithmetic stimulus materials) and for task context (whether or not the task was placed in a realistic setting).

Assessment frameworks in QUASAR also incorporated carefully delineated specifications for scoring. As was done originally in the development of the *Writing Supplement for the Iowa Tests of Basic Skills*, the QUASAR mathematics assessment developed a focused-holistic scoring protocol[11] for each task. The scoring protocols were organized with respect to three criteria for evaluating responses: mathematical knowledge, strategic knowledge, and communication. As discussed by Lane,[12] these criteria were used to develop specific scoring rubrics for each task, rubrics that at once reflected the unique mathematical demands of the task and the common framework of standards that raters were to use in scoring. The influence of this structure for task development and scoring on the technical quality of results from the QUASAR assessments is discussed below.

The careful delineation of the conceptual framework for developing the QUASAR assessment instrument given by Lane[13] provides a clear picture of the magnitude of a development effort that responds to the current demands of subject-matter and measurement specialists for high degrees of content quality in new measures of mathematics achievement. The content framework alone specifies ten cognitive processes, six content categories, six modes of stimulus presentation, and two levels of task context for a total of 720 potentially distinct types of tasks for performance assessments in mathematics. Clearly it would be absurd to propose that filling all the cells in the QUASAR content framework is necessary to constitute a content-valid and appropriate set of performance tasks for purposes of measurement. In practice, a given task likely involves many cognitive processes simultaneously, and it may display information in several modes of representation. However, the QUASAR example does serve to highlight the many aspects of a large-scale performance assessment that must be monitored to ensure fidelity to the evolving content standards of the mathematics community.

Beyond the care that should go in to the test development process to ensure content quality, it is important to recognize ways in which evolving content standards may unknowingly undermine the content validity of an assessment. Baker[14] described the difficulty of measuring certain complex skills with responses to extended performance tasks without placing perhaps undue weight on an examinee's facility with language in constructing the formal response that is the focus of evaluation. Standard 2 from the NCTM *Standards* explicitly identifies verbal components of the cognitive domain of mathematical competence. In accord with this standard, one of the QUASAR tasks asks students to study a bar graph depicting a typical day in someone's life and to respond to the graph by writing a brief story about a day in that person's life. For mathematics assessment, the variance of scores associated with linguistic factors in the evaluation of responses needs to investigated. Depending on how domain definitions from the NCTM *Standards* are made operational, this component of variance may represent a confounding factor in the use of results from extended samples of performance on complex assessment tasks. Whether or not it is considered a confounding factor, the variance associated with verbal aspects of the responses to mathematical problems is likely to loom larger than it has in more traditional approaches to measuring mathematics achievement.

Shifting definitions of content domains, not to mention the introduction of domains that are new from the standpoint of the classroom teacher, can be expected to affect certain characteristics of measures based on those domains.[15] One sees this effect in the reader and score reliabilities observed when writing samples are collected for persuasive essays in the middle elementary grades.[16] Performance of both readers and writers is much less consistent in the persuasive domain, where instructional opportunities are usually limited by a curriculum that emphasizes narrative writing. Shavelson et al.[17] illustrated the effect of domain misspecifications on estimates of score reliability in the case of hands-on science exercises. The Shavelson analyses in particular show that estimates of score reliability are likely to be markedly higher when assessment tasks are more narrowly defined and that unreliability is characteristic of poorly defined domains. As a result, inferences to broad content categories become problematic.

In the development of novel problem-solving activities to be used for assessment, researchers and practitioners will need to monitor the extent to which classroom practices keep pace with innovations in the assessment process. The development of school delivery standards[18] is a necessary part of that monitoring, but as yet no explicit guidelines have been developed for how information about opportunity to learn can be used to provide feedback for revision of assessment tasks and evaluation of technical characteristics and provide a framework for differentiated interpretations of assessment results and policy implications to enhance validity.

· ·

BEYOND PROFESSIONAL JUDGMENT IN THE VALIDATION PROCESS

The evolution of the NCTM *Standards* can be understood as a reflection of changing professional judgment about the role of mathematics education in the general cognitive development of students. The development of standards in mathematics, as well as in other parts of the school curriculum, presents a new challenge to the developers of achievement measures with respect to content quality and cognitive complexity, two aspects of the validity question discussed by Linn et al.[19] All major test publishers are presently engaged in efforts to revise instruments so that their content is more closely aligned with the NCTM *Standards*. The methods used to ascertain alignment typically involve the review of test materials by specialists in mathematics education and the classification of items according to the explicit statements of mathematics objectives.

The heavy reliance on professional judgments of content quality and, given the nature of the new mathematics standards, cognitive complexity raises critical methodological questions about this part of the validation process. The obvious question of subjectivity in the rating process can be evaluated empirically. The empirical evidence gathered so far indicates that the judgments of content experts may not be highly reliable.[20] Data that are available from content classifications of traditional test items raise questions about the fidelity of expert judges in evaluating test content. Comparisons of recent evaluations of the content of standardized achievement tests in mathematics[21] with the content specifications supplied by developers (typically determined by subject matter experts' formal analysis of content and process required to obtain correct solutions) suggest that judgments of content quality may depend heavily on the point of view of the expert making the judgments. Professional judgments, then, should not serve as the sole basis of support for or against validity in traditional testing, much less in alternative assessment procedures, without due attention to the factors that influence such judgments.

Glaser et al.[22] discuss the need for supplementing expert opinions with empirical evidence of cognitive validity of open-ended performance assessments. Such assessments, they argue, are usually developed on the basis of rational analysis and expert judgment and are *assumed* to measure higher-order reasoning because of the complexity of the tasks involved. Level of performance is often defined by psychometric difficulty and illustrative items, unaccompanied by any evidence or explanation of the underlying cognitive processes required for solutions. They also point out that what is best depicted as rational development of scoring protocols is seldom supplemented by empirical evidence indicating what knowledge and cognitive processes are actually being tested.

Potential refinements of procedures in the collection of judgment data should be considered in establishing validity of innovative assessment designs with respect to content quality and cognitive complexity. As noted by Magone et al.,[23] content analyses of tasks themselves can be expected to tell only part of the story concerning the level of cognitive complexity elicited by the task. Magone et al. used logical analyses by expert judges to validate a series of open-ended prompts designed to measure conceptual understanding, problem solving, reasoning, and communication in mathematics.

Such analyses were an integral part of the development of scoring rubrics but were also considered crucial to the validation process. However, they did not constitute the bulk of validity evidence presented to support score interpretations.

In addition to logical analyses of content, Magone et al. also examined responses to pilot and operational tasks for evidence of cognitive complexity. They coded responses to open-ended tasks in terms of solution strategy, representation and quality of written explanation, description of solution processes, and mathematical errors, arguing that empirical evidence of this sort was necessary evidence for the validity of tasks as cognitive measures of achievement in mathematics. Results were used to revise the prompts used for each task, to provide feedback for teachers about student performance, and to delete tasks from the pool to be used for later assessments.

Glaser et al.[24] selected several science performance assessment tasks for examination via student protocol analysis, including extended interviews with subjects participating in these assessments. Such analyses aim to reveal the degree of correspondence between the cognitive processes and skills the tasks were intended to measure and those actually elicited. Results of studies such as this are expected to be useful in designing more innovative approaches in which the meaning of students' scores might be more explicitly described in terms of the reasoning skills and other processes underlying their performance. Preliminary results of the Glaser et al. study suggest that the same task presented in different forms (e.g., physical manipulation of objects versus computer simulation) may elicit qualitative differences in performance.[25] Snow and Lohman[26] report other examples in which small changes in wording (e.g., abstract mathematical terms or equivalent renderings in everyday language) can have apparently large effects on cognitive processing and performance.

Interpreting the results of a content analysis of responses is not without difficulties. For example, in the Magone et al.[27] study, some tasks were discarded because responses showed little evidence of cognitive complexity. Instead, students' explanations of how they arrived at their answers used phrases such as "I used my brain," or, somewhat ironically, "logical reasoning." Although the tasks in question may have yielded no apparent evidence of "cognitively complex responses," inferring that no complex skills were required to solve the problems in question is another matter and may depend on, among other things, whether the problem was answered correctly or incorrectly.[28] In fact, the authors noted that their analysis was limited by their using written responses as the sole basis for judgments regarding the complexity of a problem or response. Like Glaser et al.,[29] they saw an important role for think-aloud protocols and interviews in understanding what is measured by open-ended tasks. To evaluate the mathematical complexity of a purportedly complex task or response clearly requires deeper analyses of the assessment process.

Unfortunately, one might question whether it is at all feasible to conduct in-depth analyses of student responses that would be useful in interpreting the results of a large-scale assessment. Time and cost considerations certainly make it impractical to interview a large number of students; even if it were practical, the mass of data collected would be extremely difficult—if not impossible—to summarize. Responses to protocol probes are often eccentric and may be interpretable only by someone who is familiar with the situation and with the student responding. On the other hand, such analyses may prove to be quite informative during task development and revision.

The most practical use for the results of such analyses with regard to score interpretation may be to point out the lack of generalizability of responses obtained from a simple content analysis. A thorough protocol analysis may reveal that similar statements correspond to radically different cognitive procedures for different students. If, for example, a student claims to have solved a problem by "using my brain," the meaning of this statement will depend on the student. One student may have had no clue as to how to solve the problem and therefore decided to disguise ignorance with an ambiguous response. Another may have understood the problem but lacked the verbal skills to describe the procedure. Still another may have been so skilled or knowledgeable that the problem was solvable almost instantaneously without any conscious awareness of the cognitive steps involved. Experts are often less able than novices to provide detailed descriptions of their problem-solving activities.[30] These concerns about the interpretability of a think-aloud protocol raise similar questions about the interpretability of written responses to probes about solution strategies in an operational assessment program.

INFLUENCES ON GENERALIZABILITY AND TRANSFER

The aspect of validity that has aroused by far the most attention from test specialists interested in performance-based alternatives to conventional achievement tests is the issue of generalizability of performance and the transfer of measured content to the unmeasured aspects of the content domain. The inferences of greatest interest to teachers, parents, and the public at large concern the broad objectives of instruction in mathematics, re-flected in the NCTM *Standards* by their emphasis on problem-solving, solution strategies, communication, and the like.

The concern about generalizability and transfer is not unique to performance-based approaches to measuring mathematical skills. In fact, these concepts occupy such a prominent place in present-day discussions of test validity because of experiences in large-scale testing programs that use conventional measures of achievement.[31] Teaching to the test poses difficulties for score interpretation not just because it compromises normative information that accompa-nies most standardized tests. It is a practice that challenges the validity of test scores as indicators of the achievement domain sampled during test construction and has been shown in high-stakes situations to distort inferences to that domain.[32] In evaluating novel approaches to assessment in mathematics, the generalizability of scores over raters, tasks, formats, and even subdomains has re-ceived considerable attention.

Influences on the variability of performance assessment tasks are observed in both the response process and the scoring process. In general, issues related to consistency in the scoring process are well understood and the overall component of score variance due to the effect of raters has been generally viewed as one aspect that, for a given performance assessment, can be easily controlled with appropriate training of raters.[33] However, the resources required to attain a given level of rater reliability are likely to vary across assess-ments for a number of reasons. Dunbar et al. noted the difference between rater reliabilities estimated under laboratory and field conditions in this regard, the former possibly giving estimates of rater consistency that are optimistic when an operational assess-ment is conducted under less than ideal conditions. Recent results from a state-level assessment of mathematics achievement based on the scoring of writing and mathematics portfolios[34] suggest possible reasons for the discrepancy between laboratory and field experiences with regard to rater reliability.

The Vermont Portfolio Assessment Program is a statewide initiative in alternative assessment unlike any other in the United States in its decentralized development of materials and emphasis on using assessment to encourage a diversity of good teaching practices.[35] It provides an exemplary operational program of the sort that is sometimes envisioned as the future of large-scale assessment, a future in which assessment tasks are determined at a local or regional level and concerns for comparability are handled through some kind of linking or calibration procedure.[36] In Vermont, writing and mathematics portfolios were assembled by fourth- and eighth-grade students from participating school districts, with "best pieces" identified. By design, there was no attempt by the state to prescribe the range of portfolio entries. In the mathematics portfolios, responses from five to seven best pieces of classroom-based tasks were scored in terms of seven scoring criteria chosen by teachers: language of mathematics, mathematical representations, presentation, understanding of task, procedures, decisions, and outcomes. The operational definitions of the four points on each of these criterion scales were determined by committees of mathematics teachers from throughout the state. Teachers also determined the method that was used to combine scores from the separate entries into a composite score for the portfolio on each criterion scale.

Koretz et al.[37] report the rater reliability coefficients for each criterion scale at each grade level. In grade four, three of the seven coefficients were below .30 (language of mathematics, outcomes, and understanding of task). The highest rater reliability coefficient (.45) was for the presentation scale. The results for grade eight were not markedly different. Although only one scale had a rater reliability below .30 (language of math), the highest coefficient (.42) was again associated with the presentation scale. The teacher-defined composite scales contained more errors due to raters than what would have been observed if a simple average or sum of scores from the separate entries had been used as the aggregate measure for individuals. Further, Koretz et al. indicate that the maximum boost to rater reliability that could be achieved with the data collected for the statewide assessment, obtained by aggregating over both portfolio entries and criterion scales, was only about .57. Such results are in stark contrast to the levels of rater reliability observed in more carefully designed and structured performance assessments.[38]

The data from the Vermont program are revealing for several reasons. First, they reinforce the remarks made earlier about rater reliability in field settings. Second, they suggest that the teachers on the reading committees either had not reached consensus regarding the definitions of score points for each criterion scale or perhaps had not had enough experience with the criterion definitions to make consistent judgments about how levels would be revealed in actual student responses. Third, they show that the task of keeping straight four score points on seven different criterion scales (28 scale points in all) may be too demanding for even experienced and motivated teacher-raters. Finally, the Vermont results exemplify some of the procedural difficulties of using essentially unstructured tasks for large-scale assessment purposes. The only structure built into the assessment design involves the specification of best pieces on the response end and explicitly defined criteria on the rating end.

Another aspect of the Vermont data that deserves comment involves the frequency distributions of ratings, which showed an extremely high concentration of ratings at one or two points on the scale for several criteria. For such criteria the reliability coefficients demonstrate the usual effects of restriction of range on the ratings, and they are somewhat difficult to interpret as a result. Koretz et al.[39] note this effect. The fact that for some scales it was revealed as a strong floor effect on the ratings—92 percent of the grade 4 sample received the lowest possible rating 1 on the outcomes scale—whereas for others the concentration was in the middle of the distribution raises doubts about the quality of the anchor points across the seven criterion scales. The authors argue that explaining low rater reliability in terms of statistical artifacts such as attenuation due to range restriction does not answer the more important question of what caused the reduction in variability in the first place.

Although early studies of performance-based assessment concentrated on raters in the estimation of components of score variance, recent studies of the generalizability of extended responses to complex tasks also have raised fundamental questions about the behavior of examinees during the response process. Linn[40] observed that "high levels of generalizability across tasks alone *do not guarantee* valid inferences or justify particular uses of assessment results. But low levels of generalizability across tasks . . . pose serious problems regarding comparability and fairness to individuals who are judged against performance standards based on a small number, and perhaps different set, of tasks."[41] Several recent examples of the generalizability of performance-based tasks in mathematics characterize the challenges faced by designers of innovative large-scale assessments.

Lane et al.[42] reported on the reliability and validity of the QUASAR Cognitive Assessment Instrument (QCAI), a set of open-ended tasks measuring mathematical reasoning skills of middle-school students. Conscientious selection and training of raters resulted in high interrater reliability. Variation in scores due to choice of rater or to any interaction between rater and student or rater and task was negligible. However, substantial variation due to student-by-task interaction (between 55 and 68 percent of total variance) was revealed in the generalizability study. Scores were found to depend to a substantial degree on the particular set of items administered, and the authors indicate that clear inferences about a student's mathematical reasoning ability were uncertain. The Lane et al. results are particularly relevant because their tasks and scoring rubrics were developed with considerable attention to the NCTM *Standards*.

Despite the large variance component due to person-by-task interaction, the overall generalizability coefficients for the nine tasks included on a given form of the assessment were in the .7 to .8 range. These values are markedly higher that many generalizability coefficients reported in the performance assessment literature, and they suggest a principle for performance assessment design that has long been recognized in the development of conventional achievement tests, namely that high levels of person-by-task interaction can be tolerated as long as the number of tasks (items) in the assessment is sufficiently large. The addition of a few more tasks in the QUASAR assessment would bring the generalizability coefficients into the range that is typical for objective tests of mathematics achievement (each performance task is worth roughly two multiple-choice items from the standpoint of reliability).

There is a growing body of evidence collected from studies of performance tasks in a variety of content domains that a substantial number of open-ended tasks may be required to support high-stakes uses of results for decisions about individuals.[43] To offer specific guidelines for the number of tasks needed to secure a reliability coefficient of, say, .90 can be quite misleading because of the many factors (quality of raters, parallelism of additional tasks, delineation of the content domain, etc.) known to influence such assessments. As noted by Ruiz-Primo et al.[44] in their study of hands-on science performance exercises, "increasing the number of occasions would increase the [generalizability] coefficients (four occasions to achieve reliability .80), but it would do so at considerable cost."[45]

Another example of the use of open-ended tasks in large-scale mathematics assessment provides another perspective for understanding the nature of information about achievement in mathematics that is obtained by innovative approaches. Stevenson et al.[46] describe the characteristics of open-ended geometry proofs administered to more than 43,000 high school students in North Carolina. In contrast to the Vermont results, reader reliabilities based on two independent readings of the same task, rated by a focused holistic approach, were above .90 for all proofs included in the assessment. Either through content definitions or training practices, geometry proofs were clearly amenable to the rating process in a way that the unstructured portfolios in Vermont were not.

The generalizability question was addressed by Stevenson et al. by the inclusion of a subsample of students who also took a multiple-choice proofs test in addition to the open-ended version. The disattenuated correlations between the multiple-choice proofs test and the open-ended problems were quite high (approximately .85). The disattenuated correlation between an individual open-ended proof and the same proof in multiple-choice format would be estimated to be nearly perfect (.99, assuming the reliability of the multiple-choice proofs test to be around .80). From these results, it is clear that performance on what are in this case highly structured, open-ended tasks in geometry does transfer to performance in a traditional format. The irony of this example is that the degree of transfer appears to be so high as to beg the question about whether the formats are measuring anything different about achievement in the relevant domain. That the formats send different messages to the audiences of an assessment about what is valued in the geometry curriculum, of course, is a separate issue.

Given the results of the North Carolina study, a relevant question for the development of mathematics performance tasks and rating scales might be phrased as follows: When the factors that produce lack of generalizability on complex tasks in mathematics are controlled to a degree deemed necessary for large-scale applications, will the constructs measured by the tasks rank-order examinees any differently than would a conventional test of related mathematics skills?

The influence of problem format on variability in responses is an obvious consideration in understanding the levels of score reliability or generalizability that have been observed in open-ended assessments of mathematics achievement. Although the influence of format on difficulty was noted earlier, the effects of format can be quite subtle and relate more to the construct interpretations that are made of scores. Webb and Yasui[47] examined the performance of seventh graders on three types of open-ended mathematics items: unembellished numerical exercises; short, one-question, word problems; and relatively complex, extended word problems. Among the three types of problems, no differences in difficulty were found; students displayed similar performance regardless of the item format. Students were especially variable in terms of their ability to set up the problem correctly regardless of the amount of verbal context supplied.

What differences were apparent across item types concerned the kinds of errors students made. Computational errors occurred less frequently for verbal items, perhaps because the context provided clues enabling students to check the reasonableness of their answers. Extended word problems elicited more misinterpretations of the question and uninterpretable responses and omitted more answers than did the shorter verbal items. The authors suggested that cognitive overload, frustration, and abated motivation may affect performance on these lengthy, "realistic" problems.

Each of the extended problems was presented in several parts or subquestions; many students failed to see the connection between the parts and recomputed figures already obtained in a previous subquestion. It was not clear to the researchers to what extent "erosion of performance"[48] on the extended word problems revealed a weakness in *mathematical* skills. Student errors may have reflected, for example, difficulty in reading or interpreting lengthy verbal passages, poor skills in organizing verbal information, or an inability to sustain interest and motivation without a new stimulus. Failure to build on previous subquestions could have been the result of previous experience with mathematics tests in which each item presents a problem unrelated to the questions preceding it.

Webb and Yasui demonstrate the substantive importance of understanding the reasons for lack of generalizability in the context of extended samples of student performance in mathematics. In so doing they reveal an important connection in the validity argument between generalizability analyses and construct interpretations of the results of a mathematics assessment. The extent to which performance transfers from one complex task to another, or perhaps from sets of related tasks to other sets, in one sense determines the specificity of the domain to which defensible inferences can be made from the results of an assessment. Evidence of poor transfer does not necessarily undermine the value of an assessment for a given purpose. However, poor transfer severely restricts the range of legitimate uses of assessment information and exacerbates the problem of how to communicate results to audiences who have developed high expectations regarding the utility of the enterprise.

. . . .

TASK SAMPLING AND AGGREGATE REPORTS TO THE PUBLIC

How many tasks will be needed from new forms of assessment to secure a valid measure of achievement for a particular purpose, and how will results of locally controlled assessment programs using novel tasks and formats be combined for use in policy discussions at state and national levels? The recent report to Congress by the Office of Technology Assessment (OTA) [49] reflects on the shared experience of nearly all students when the essay question just happened to cover what had been studied the night before, as well as "the time they 'bombed' on a test, unjustly they felt, because the essays covered areas they had not understood or studied well." [50] The OTA report also addresses the concerns that arise when aggregate results are of most interest.

The issue of task sampling presents exactly the kind of ill-structured problem that has no generic right answer but instead has many potentially right answers depending on the circumstances of test use. A broadly defined content domain such as problem solving might require extensive sampling of tasks because of the influence context can have on problem-solving behavior.[51] As guideposts for the development of new kinds of assessment instruments, the NCTM *Standards* tend to emphasize the importance of similar broad domains of mathematical competence. Unfortunately, there is limited empirical evidence from experimental measures in such domains of the kinds of generalizability that might be expected and, hence, little empirical basis for recommendations concerning the number of tasks that might be necessary for a given use of results. What is known about content sampling from the standpoint of conventional achievement tests provides clear evidence that the meaning of a test score can be quite easily manipulated by purposeful selection of items to match the objectives of a local curriculum or policy initiative.[52]

Whenever there is a general concern about the sampling of tasks, there is a concomitant concern over the possibility that influences on task performance will be concentrated in subpopulations of examinees—subpopulations differentiated by race, gender, or some other correlate of opportunity to learn.[53] On the subject of differential functioning of test questions by group, some specialists go so far as to argue there is no such thing as an unbiased item; rather, the responsibility of test developers is to ensure that content domains are sampled in such a way to balance out the bias, that is, to include enough variety in stimulus materials and balance in content that the assessment as a whole does not systematically favor one group over another. It is perhaps this aspect of alternative measures of achievement in mathematics that the research community knows the least about. Understanding the nature of group differences on novel measures of performance is also an aspect of instrument development that may have the greatest impact on the consequential validity of the next generation of assessments.[54] The fact that new assessments of achievement in mathematics are likely to focus on new aspects of the mathematics curriculum identified by the NCTM *Standards* makes the monitoring of shifts in teaching practice critical to the valid use of results.

In addition to the extremely limited data available on differential item functioning (DIF) of performance task with respect to gender or ethnicity, there is also a limited understanding of how to detect the phenomenon. Recently proposed methods by Welch and Hoover[55] and Miller and Spray[56] are in their infancy when compared with methods developed for dichotomously scored test items. In part because of the absence of suitable methodology, there are virtually no systematic investigations of DIF in the performance assessment literature in any content domain. As data from large-scale performance-based measures of mathematics achievement become more readily available, the issue of differential task functioning will need to be carefully evaluated.

Combining the results of novel assessments across disparate tasks, geographic regions, school district boundaries, or other demographic groupings is of great interest to policymakers intent on using the results of new forms of assessment to monitor educational reform efforts. Linn [57] distinguishes statistical and judgmental approaches to the process of linking the results of distinct assessments, and describes a continuum of inferences that might be justified depending on the conditions under which the link across assessments was established. Generally speaking, the kinds of assessments that are currently being proposed in the context of educational reform efforts can only be linked across sites through some form of calibration, but calibration by professional judgment.

One empirical example of an attempt to link direct writing assessments across states was described by Linn et al.[58] essays written by students from one state were evaluated with the scoring protocols from another state. Linn et al. found a surprising degree of consistency in the way students were rank-ordered by the panels of readers from different states. However, absolute judgments about the level of performance reflected by an essay response were quite different, revealing that readers from one state did not share the same standards for performance as did readers from another state. The implications of this finding for large-scale decentralized implementations of performance-based approaches to assessment are profound given the usual impetus for such programs, the maintenance of standards.[59]

Linn[60] discusses the use of what he calls "social moderation" to link the results of distinct assessments. Social moderation seeks to develop consensus among educators about standards for performance and exemplars of those standards and relies on such shared understanding to provide the link that statistical procedures do for more conventional assessments. As was implied previously, common understandings about a content domain as shared standards for performance do evolve over time. The extent to which social moderation can be relied on for the calibration of scores on innovative assessments in mathematics, however, is a matter of speculation. At present, no procedure for linking disparate assessments in mathematics can be recommended *because of the good data it has produced* to date. The Vermont results are sobering in this regard and compel us as researchers to inquire further into the methods that will be needed if a decentralized approach to the assessment process is to yield broad inferences to direct state and national education policy through the linking of locally developed instruments.

SUMMARY AND CONCLUSIONS

In exploring innovations in large-scale measures of achievement in mathematics, it is important to recognize that a diversity of assessment procedures applied in a diversity of situations precludes the production of sweeping, general statements about various types of assessments and the technical issues involved in implementing them. Traditional concerns with reliability and validity are expanded to encompass issues raised by a host of related criteria.

The domain and definition of mathematics has been expanded in the NCTM *Standards* to include skills containing components that might have previously been classified as nonmathematical, perhaps verbal or analytical. This expanded domain creates new problems for assessment. How can verbal skills, for example, that are part of the new mathematics domain be measured without confusing their measurement with that of verbal skills that are separate from mathematics achievement? The broadening of the domain also makes it more difficult to achieve highly reliable measurement of mathematical skills.

The classification of assessment tasks according to content and complexity relies heavily on the judgment of content experts. Consensus among experts has often proved more difficult to attain than one might expect. Evidence of content validity provided by professional judgments needs to be supplemented with empirical evidence of cognitive validity. The logistics of collecting such evidence have not been fully researched; the methods used may themselves present new problems. In-depth protocol analyses, for example, are impractical to conduct on a large scale, and the mass of data they produce may be difficult to interpret and impossible to synthesize for use in the interpretation of aggregate reports of mathematics achievement.

With a variety of alternative assessment procedures, issues of generalizability and transfer necessarily become more complicated. Inferences to the domain can still be hampered by teaching to the test, although the practice may take different forms when the assessment instrument is not a traditional test. Issues related to generalizing over raters (scoring reliability) are relatively well understood. Nevertheless, raters do not always behave as expected. When well-trained raters fail to score tasks consistently, as in the case of the Vermont Portfolio Assessment Program, clearly there are aspects of the scoring process that are not yet fully understood. A deeper understanding of the scoring process is essential if we wish to avoid wasting precious resources on large-scale assessments that produce nearly useless results.

Evidence obtained thus far indicates that generalizability across tasks is often low. Students may vary greatly in their performance on mathematics assessments depending on the particular tasks by which they are tested. Apparently some aspects of performance in mathematics are highly subdomain-specific or require specified knowledge above and beyond transferable skills. The inferences that can be made about the scope of students' mathematical abilities are severely limited unless evidence of across-task generalizability can be obtained. Further research is needed to understand better the factors affecting this aspect of generalizability. Short of such understanding, broad sampling of assessment tasks from specified content frameworks will be needed to support the broad inferences to domains of mathematical achievement that have caught the fancy of education and public policymakers.

Although there is evidence that the use of different formats to assess the same knowledge and skills may have little effect on students' level of performance, the types of errors students make (and therefore perhaps the cognitive skills being tested) may depend on the format in which an assessment task is presented. This is the thrust of the argument in favor of new formats for tests; however, newly developed item types must be studied to determine whether the skills they measure differ from those assessed by other formats, as well as what potential advantages and pitfalls they may present.

Like the 1890s, the 1990s are characterized by optimistic expressions of faith in new educational standards. According to the broad brush of optimism, by setting standards high and holding all students to them, leaving none behind in Hamlin, educational leaders expect to see tomorrow's students stride into adulthood fully prepared for the demands of life and work in the twenty-first century. As the last century saw the rise of new assessment procedures to support the maintenance of the educational standards of that time, so now a proliferation of new, innovative assessments is already arising to measure progress in meeting today's new standards. The literature is full of optimistic statements about the purported advantages of these new procedures. Surely optimism is good, yet naive optimism can be treacherous. It is imperative to recognize that new assessments may bring new problems. Long periods of debugging and refining new procedures may be required before alternative assessments can produce results that are meaningful and widely applicable. Moreover, it would be naive to expect that new procedures will be less amenable to abuse than traditional measures have been. Disregard for individual differences, the exclusionary use of assessment results, various forms of teaching to the test, and other undesirable outcomes are as likely to occur with today's alternative assessments as with traditional instruments.

Unlike the standards movement of the last centurial transition, many of these outcomes are not unanticipated and can be guarded against. If we can curb our optimism long enough to examine the efficacy, challenges, and potential consequences of new assessment procedures before they are implemented for high-stakes purposes, we can avoid possible negative outcomes, invest more time and resources in positive refinements, and ultimately produce better, more useful measures of achievement in mathematics.

ENDNOTES

1 M. McConn, "The uses and abuses of examinations," in *The Construction and Use of Achievement Examinations*, ed. H. E. Hawkes, E. F. Lindquist and C. R. Mann, (Boston: Houghton Mifflin Company, 1936).

2 Ibid., 447. Emphasis in the original.

3 R. L. Linn, E. L. Baker, and S. B. Dunbar, "Complex performance-based assessments: Expectations and validation criteria," *Educational Researcher, 20*, (1991), 15-21.

4 Ibid.

5 S. Messick, "The Interplay Between Evidence and Consequences in the Validation of Performance Assessments" (Paper presented at the annual meeting of the National Council on Measurement in Education, San Francisco, April, 1992).

6 S. B. Dunbar, D. M. Koretz, and H. D. Hoover, "Quality control in the development and use of performance assessments," *Applied Measurement in Education, 4*, (1991) 289-304.

7 Cf. A. N. Hieronymus and H. D. Hoover, *Iowa tests of basic skills: Writing supplement teacher's guide* (Chicago: Riverside, 1987); A. N. Applebee, J. A. Langer, and I. V. S. Mullis, *Writing: Trends across the decade, 1974-84* (National Assessment of Educational Progress Rep. No. 15-W-01) (Princeton, NJ: Educational Testing Service, 1986).

8 National Council of Teachers of Mathematics, *Curriculum and evaluation standards for school mathematics*, (Reston, VA: Author, 1989).

9 S. Lane, "The conceptual framework for the development of a mathematics performance assessment instrument," *Educational Measurement: Issues and Practice, 12*, (1993), 16-23.

10 Ibid.

11 *Iowa tests of basic skills.*

12 "Conceptual framework."

13 Ibid.

14 E. L. Baker, *The role of domain specifications in improving the technical quality of performance assessment*, Technical Report, (Los Angeles: Center for Research on Evaluation, Standards, and Student Testing, 1992).

15 Baker, *Domain specifications*; R. J. Shavelson, X. Gao, and G. P. Baxter, *Content validity of performance assessments: Centrality of domain specification*, Technical Report, (Los Angeles: Center for Research on Evaluation, Standards, and Student Testing, 1992).

16 *Iowa tests of basic skills.*

17 *Content validity.*

18 National Council on Education Standards and Testing [NCEST], *Raising standards for American education*, (Washington, DC: United States Congress, 1992).

19 "Complex performance-based assessments."

[20] W. J. Popham, "Appropriate expectations for content judgments regarding teacher licensure tests," *Applied Measurement in Education, 5*, (1992), 285-302.

[21] Cf. G. F. Madaus et al., *The Influence of Testing on Teaching Math and Science in Grades 4-12: Executive Summary.* (Boston, MA: Boston College, Center for the Study of Testing, Evaluation, and Educational Policy, 1992) 1; T. A. Romberg and L. D. Wilson, "Alignment of tests with the standards," *Arithmetic Teacher* (40) 1992, 18-22.

[22] R. Glaser, K. Raghavan, and G. P. Baxter, *Cognitive theory as the basis for design of innovative assessment: Design characteristics of science assessments,* Technical Report, (Los Angeles: Center for Research on Evaluation, Standards, and Student Testing, 1992).

[23] M. Magone, J. Cai, E. A. Silver, and N. Wang, "Validity evidence for cognitive complexity of performance assessments: An analysis of selected QUASAR tasks," *International Journal of Educational Research*, in press.

[24] *Cognitive theory.*

[25] G. P. Baxter, *Exchangeability of science performance assessments,* (Unpublished doctoral dissertation, University of California, Santa Barbara, 1991).

[26] R. E. Snow and D. L. Lohman, "Implications of cognitive psychology for educational measurement," in *Educational Measurement*, 3rd ed., ed. R. L. Linn, (New York: Macmillan, 1989), 263-331.

[27] Magone et al., "Validity evidence for cognitive complexity."

[28] "Implications of cognitive psychology."

[29] *Cognitive theory.*

[30] "Implications of cognitive psychology."

[31] R. L. Linn, M. E. Graue, and N. M. Sanders, "Comparing state and district test results to national norms: The validity of the claims that 'Everyone is above average'," *Educational Measurement: Issues and Practices, 9*, (1990), 5-14; L. A. Shepard, "Inflated test score gains: Is the problem old norms or teaching to the test?" *Educational Measurement: Issues and Practices, 9*, (1990), 15-22.

[32] D. M. Koretz, R. L. Linn, S. B. Dunbar and L. A. Shepard, "The effects of high-stakes testing on achievement: Preliminary findings about generalization across tests," (Paper presented at the Annual Meeting of the American Educational Research Association, Chicago, April, 1991).

[33] "Quality control."

[34] D. M. Koretz, D. McCaffrey, S. Klein, R. Bell, and B. Stecher, *The reliability of scores from the 1992 Vermont portfolio assessment program,* Technical Report, (Washington, DC: The RAND Corporation. 1992).

[35] Ibid.

[36] Cf. R. L. Linn, "Linking results of distinct assessments," *Applied Measurement in Education, 6*, (1993), 83-102.

[37] *Reliability of Vermont scores.*

[37]

[38] Cf. Dunbar et al., "Quality control"; S. Lane, C. A. Stone, R. D. Ankenmann, and M. Liu, "Empirical evidence for the reliability and validity of performance assessments," *International Journal of Educational Research*, in press).

[39] *Reliability of Vermont scores.*

[40] R. L. Linn, "Educational assessment: Expanded expectations and challenges," *Educational Evaluation and Policy Analysis*, *15*, (1993) 1-16.

[41] Ibid., 27. Emphasis in the original.

[42] "Empirical evidence."

[43] "Educational assessment."

[44] M. A. Ruiz-Primo, G. P. Baxter, and R. J. Shavelson, "On the stability of performance assessments," *Journal of Educational Measurement*, 30, (1993), 41-54.

[45] Ibid., 46.

[46] Z. Stevenson, C. P. Averett, and D. Vickers, "The reliability of using a focused-holistic scoring approach to measure student performance on a geometry proof," (Paper presented at the Annual Meeting of the American Education Research Association, Boston, April, 1990).

[47] N. Webb and E. Yasui, *Alternative approaches to assessment in mathematics and science: The influence of problem context on mathematics performance*, Technical Report, (Los Angeles: Center for Research on Evaluation, Standards, and Student Testing, 1991).

[48] Ibid, 23.

[49] Office of Technology Assessment, *Testing in American Schools: Asking the Right Questions*, (Washington DC: United States Congress, 1992).

[50] Ibid., 242.

[51] J. H. Larkin, "What kind of knowledge transfers?" in *Knowing, learning, and instruction: Essays in honor of Robert Glaser*, ed. L. B. Resnick, (Hillsdale, NJ: Erlbaum, 1989).

[52] R. L. Linn, and R. Hambleton, "Customized Tests and Customized Norms," *Applied Measurement in Education*, 4, (1991), 185-207.

[53] L. Feinberg, "Multiple choice and its critics," *The College Board Review*, No. 157. (1990); Linn et al., "Complex performance-based assessments"; S. B. Dunbar, "Comparability of indirect measures of writing skill as predictors of writing performance across demographic groups," (Paper presented at the annual meeting of the American Educational Research Association, Washington, D.C., April, 1987).

[54] "Evidence and consequences."

[55] C. Welch and H. D. Hoover, "Procedures for extending item bias detection techniques to polytomously scored items," *Applied Measurement in Education*, 6, (1993), 1-19.

[56] T. R. Miller and J. A. Spray, "Logistic discriminant function analysis for DIF identification of polytomously scored items," *Journal of Educational Measurement*, 30, (1993) 107-122.

[57] "Linking results of distinct assessments."

[58] R. L. Linn, V. L. Kiplinger, C. W. Chapman, and P. G. LeMahieu, "Cross-state comparability of judgments of student writing: Results from the New Standards Project," *Applied Measurement in Education*, 5, (1992), 89-110.

[59] Cf. National Council on Education Standards and Testing, *Raising standards for American education*.

[60] "Linking results of distinct assessments."

Legal and Ethical Issues in Mathematics Assessment

Diana C. Pullin[1]
Boston College

After two decades of efforts at the state and local levels to reform the nation's elementary and secondary schools, the United States moves toward the millennium firmly committed to a series of national reform initiatives. These national efforts consist of two types of often intersecting approaches to driving enhanced educational productivity. First, at the urging of the nation's governors, the federal government has initiated a series of efforts to promote national curriculum, performance, and opportunity to learn standards that would be adopted by states and local school districts on a voluntary basis. These federal initiatives seek to promote systemic state reform by creating a new federal role developing national education standards and assessments and setting benchmarks to measure progress toward attaining those goals. Although no federal mandates or sanctions are proposed, federal seed money and other financial aid, coupled with technical assistance will have a significant impact. Moreover, the power of the federal government to lead a public forum addressing the goals will also mean a strong central vision for the reforms.

At the same time as these federal initiatives are being pursued through President Bush's America 2000 initiative and President Clinton's Goals 2000: Educate America Act, several professional organizations have been pursuing similar objectives. For example, the National Council of Teachers of Mathematics (NCTM) and other professional associations have defined curricular content standards or frameworks, benchmarks, and performance standards in their subject areas. Efforts such as these mark an important change within the field of education: systematic, political reform driven from within the profession rather than externally mandated change compelled by some governmental entity.

Current efforts at educational reform include the work of the National Research Council's Mathematical Sciences Education Board (MSEB), which has proposed that assessments be used to meet three goals. First, assessment should be used to support or improve teaching of important mathematics content and procedures. Second, mathematics assessment should support good instructional practice. Finally, assessment should support every student's opportunity to learn important mathematics.

MSEB's proposals to enhance educational achievement in mathematics and to increase access to educational opportunities to learn can be evaluated as part of a movement that has provoked public debate and scrutiny of our schools for over 40 years. Past efforts, particularly federal ones, to mandate equality of educational opportunity through laws, regulations, and court decisions will impact the current reform proposals, even if they are characterized as "voluntary", rather than mandatory, and "national" rather than federal. The proposed use of assessment as a tool of educational reform prompts comparison with prior efforts to enhance educational achievement with high stakes tests which have significant consequences for individual test-takers. It has been in these "high stakes" situations in which the legal impact of these types of initiatives has been most complicated.

In the years since the decision of the U. S. Supreme Court in *Brown* v. *Board of Education*,[2] which found a constitutional bar to state laws segregating schools on the basis of race, there has been a large increase in the use of state and federal statutes, regulations, and court decisions to regulate educational practices and educational testing efforts. Judges and lawmakers have scrutinized educational reform proposals or imposed legal mandates upon educational practices in efforts to attain desirable social, political and educational goals, especially our commitment to equity and fairness.

Few would dispute the desirability of a dramatic increase in educational achievement of the nation's students. However, part of the discussion of reform initiatives must focus on whether these changes can be implemented without undermining the nation's long-standing commitment to equity. These questions are rooted largely in consideration of both public policy and the law, because the law is frequently used to address efforts to maintain our traditional commitments to fair treatment of all and our aspiration to educate all children well. Although not providing answers to all of these questions, this paper will attempt to highlight the issues that should be considered by those participating in this debate.

RACE AND THE EFFECTS OF EDUCATIONAL REFORM

A precise legal analysis of assessment reform proposals based on MSEB's new principles for mathematics assessment will depend upon how those proposals are implemented. However, MSEB's dual goals of using assessment to enhance mathematics learning and promoting equity by supporting every student's opportunity to learn parallel some earlier initiatives that were subject to keen policy debates and intense legal scrutiny.

An earlier, and similar educational reform effort involved the use of minimum competency tests (MCT) to determine whether a student would receive a high school diploma. The analogy is useful if a national system of tests or assessments or what might become, de facto, a national exam results from these proposals or if a school or state or local education agency were to use mathematics assessments to determine the award of diplomas, proficiency certificates, or access to employment. Constitutional scrutiny of any educational program, as was the situation with the state or local MCT programs reviewed previously by the courts, can be triggered whenever any governmental body, be it federal, state or local, acts on a voluntary basis to sort people into groups for differential treatment. The level of this scrutiny intensifies as the stakes attached to the groupings go up; if the stakes attached to a mathematics assessment are high enough to involve high school diploma denial, limitations on access to particular curricular tracks, to higher education or the workplace, or stigmatizing labels for individuals who do not succeed on the assessment, then the possibility for successful legal challenges to the program increases.

Perhaps the most noteworthy legal review of an educational reform initiative with a high stakes use of testing was *Debra P. v. Turlington*, a challenge to the state of Florida's program to condition the award of a high school diploma upon successful performance on a minimum competency test.[3] Florida's legislative goals were to promote educational accountability and insure that every school district provided "instructional programs which meet minimum performance standards compatible with the state's plan for education. . . [and] information to the public about the performance of the Florida system of public education in meeting established goals and providing effective, meaningful and relevant educational experiences designed to give students at least the minimum skills necessary to function and survive in today's society".[4] The test used to measure student performance in reading, writing, and mathematics was commonly known as the Functional Literacy Test. Initial failure rates on the test were high and a disproportionate number of those failing the test were black; the early failure rate among black students was approximately ten times that for white students.[5]

The racial impact of Florida's test required the courts to assess the program under Title VI of the Civil Rights Act of 1964[6] and the U.S. Constitution. In reviewing constitutional challenges to the program presented by students of all races who failed the test, the courts that decided *Debra P.* addressed validity and reliability issues in educational testing. In deciding the case, the courts looked, by analogy, to standards used in reviewing employment testing under Title VII of the Civil Rights Act of 1964[7] and to a series of teacher testing cases brought under the U.S. Constitution.[8] In *Debra P.*, the Fifth Circuit held that a state "may not constitutionally so deprive its students [of a high school diploma based upon test performance] unless it has submitted proof of the curricular validity of the test."[9] The court further explained that "if the test covers material not taught the students, it is unfair and violates the Equal Protection and Due Process clauses of the U.S. Constitution."[10] The constitutional protections were triggered because of the magnitude of the consequences of test failure, that is, denial of the diploma, restrictions on access to employment and higher education, disproportionate impact on minority students, and the stigmatizing effect of being regarded as "functionally illiterate."

In considering test performance in which black students consistently fail at rates far higher than whites, *Debra P.* looked at issues of test bias and also held that Title VI would require a demonstration that the test was a fair test of what was taught in schools[11] and that the government had taken steps to eliminate the effects of past unlawful racial discrimination that might impact test performance.[12] Further, "in attempting to justify the use of an examination having. . . a disproportionate impact upon one race. . . [the government must] demonstrate either that the disproportionate failure of blacks was not due to the present effects of past intentional segregation or, that as presently used, the diploma sanction was necessary to remedy those effects."[13]

Some of the same types of race discrimination identified in the *Debra P.* situation could occur in a mathematics assessment system. Any assessment tasks that require knowledge that might not be taught in school or might not be part of a common cultural norm could negatively impact performance on the basis of race, ethnicity, gender, or socioeconomic status.[14] For example, a proposal by the 1991 Victorian (Australia) Curriculum and Assessment Board promotes the use of simulation to assess performance. One of their proposed simulations asks students to investigate "the chance of winning a tennis game after being two sets down." The success of former African American tennis pro Arthur Ashe aside, few minority youngsters would have a fair chance at succeeding on this task.[15] Similarly, the use of a projects-based approach to mathematics assessment could have a deleterious effect on low-income or limited-English-proficient youngsters if a significant portion of the assignment was to be done as homework where parental assistance might play a role in successful project completion.

The courts have also invalidated the use of testing and instructional practices which resulted in dead-end educational tracking for low-performing students. In the *McNeil* v. *Tate* case,[16] the use of class assignment practices which resulted in segregation of African American students in low track placements which they never left and which provided limited opportunities for educational achievement were declared unlawful.[17] Similarly, in *Larry P.* v. *Riles*,[18] federal courts banned the use of intelligence tests to place black students in special education classes in California due to undesirable consequences for these students. Any use of mathematics assessments to place students in educational tracks will probably not withstand such legal review unless educators can demonstrate that students placed in low tracks have the opportunity for greater educational achievement as a result of the tracking.

One should assume, at least during the early period of implementation of the project, that disproportionate numbers of minority students will not succeed on the assessments, whatever form they take. If as a consequence, minority students are denied competency certificates, diplomas or access to greater educational opportunity at a rate higher than their proportion of the total population of students being assessed, then significant legal problems could ensue.

Other legal standards apply if there is any effort to link education and the workplace through the use of assessments of mathematics skills in order to certify proficiency to employers. There is currently widespread interest among politicians, business leaders, and some educators to create these closer links between schools and the workplace. Such initiatives could trigger legal provisions concerning protections against discrimination in employment on the basis of race, ethnicity, national origin, gender, and handicapping condition.[19]

As one example of the effort to link schools and the workplace, a group of educational, business, and labor leaders participated in the U.S. Labor Secretary's Commission on Achieving Necessary Skills (SCANS) which seeks to fulfill the mission set forth in President Bush's America 2000 initiative and "to establish job-related. . . skills standards, built around core proficiencies."[20] SCANS determined that proficiency levels ought to be defined at several levels, from preparatory, to work-ready, to intermediate, to advanced, to specialist.[21] SCANS' perceptions of proficiency are extremely ambitious. For example, for mathematics and computational skills, SCANS concludes that "virtually all employees should be prepared to maintain records, estimate results, use spread sheets, or apply statistical process controls if they negotiate, identify trends, or suggest new courses of action."[22] SCANS estimated that "less than half of young adults can demonstrate the SCANS reading and writing minimums; even fewer can handle the mathematics."[23] To the extent that mathematics assessments track the SCANS' goals, there may be both education and employment-related legal consequences attached to mathematics reform initiatives. If a mathematics assessment has the impact of determining an individual's access to employ-ment then, for purposes of legal and public policy analyses, the assessment may become, in essence, an employment test.

Much legal scrutiny has been applied to the use of tests or assessments' both formal and informal, to determine the employ-ment opportunities of racial and ethnic minorities and underrepresented men or women in a workforce. Title VII of the Civil Rights Act of 1964 bars discrimination in employment,[24] but allows employers to use "professionally developed ability tests. . . [which are not] designed, intended, or used to discriminate."[25] These provisions have been interpreted by the Supreme Court in *Griggs* v. *Duke Power* as a bar to the use of employment tests that have an adverse impact on protected groups unless the employer can establish that the test ". . . bear[s] a demonstrable relationship to successful performance of the jobs for which it [is] used." [26] This interpretation of Title VII was ratified by the Congress when it enacted the Equal Employment Opportunity Act of 1972 (P.L. 92-261).[27] Further, the Supreme Court has held that "Title VII forbids the use of employment tests that are discriminatory in effect unless the employer meets 'the burden of showing that any given require-ment [has]. . . a manifest relation to the employment in question' [and showing] that other tests or selection devices, without a similarly undesirable racial effect, would also serve the employer's legitimate interest in 'efficient and trustworthy workmanship.' " [28] It is important to note that the types of tests and criteria struck down by the Court under these "job relatedness" standards have included general high school diploma requirements and standardized tests of general ability, such as the Wonderlic.[29] Discriminatory tests have been found impermissible

> . . . unless shown, by professionally acceptable methods, to be predictive of or significantly correlated with important elements of work behavior which comprise or are relevant to the job or jobs for which candidates are being evaluated. [30]

In pursuing the inquiry required by this legal standard, the courts have delved deeply into the technical details of employers' validity and reliability studies,[31] including the accuracy of job analyses for determining content validity.[32]

The Supreme Court in Griggs also noted that performance tests should be keyed to higher level jobs in an employment context only when an employer can demonstrate that "new employees will probably, within a reasonable period of time and in a great majority of cases, progress to a higher level."[33] Further, where disparate results occur, differential validation should be done on minority and white groups whenever technically feasible.[34]

These technical and legal standards would, in a traditional Title VII case, be imposed on employers. However, given the potential for an interrelationship between educational institutions and employers who might wish to rely on individual assessment data, there is the possibility of new types of scrutiny of educational practices if mathematics assessment is linked to employment opportunities.

An irony associated with the current national proposals being discussed is that the effort to move to *uniform* competency standards for all that are beyond the minimum increases the potential for difficulty for an employer (or a school) attempting to defend the standards in a job relatedness inquiry under Title VII. The more that the skills move from minimal, basic standards, the harder it will be to establish the business necessity of performance of *all* of the skills in *all* jobs in a particular workplace (or, to set a somewhat lower goal, for *all* jobs to which all employees in a workplace might realistically aspire).[35]

The effect of the Civil Rights Act of 1991[36] has been both a clarification of the standards for assessing discrimination in employment and a strengthening of the legal remedies for intentional, unlawful discrimination. The act codifies a long history of U.S. Supreme Court decisions since *Griggs* v. *Duke Power*[37] defining the "business necessity" defense for discriminatory acts in employment and the "job relatedness" requirement for employment requirements. The Act also bars the practice used in some employment testing programs of statistically adjusting or using different cutoff scores on the basis of race, color, religion, sex, or national origin.[38]

Many of those currently promoting the use of assessment to enhance educational achievement and the infusion of workplace-related skills into the assessments have proposed the use of assessment data to determine such things as the award of certificates of mastery of workplace skills. Under such proposals, schools would award competency certificates and employers would use them to make hiring, placement, and promotion decisions. This approach opens the assessments to challenge as employment tests and, as such, could subject them to the "business necessity" or "job-relatedness" standards set forth by the courts if there is a disproportionate impact on the groups protected by Title VII. As consideration of the implementation of assessment reform proceeds, caution should be exercised concerning these school-to-workplace linkages in the use of mathematics assessments.

IMPACT ON PEOPLE WITH HANDICAPPING CONDITIONS

Through both the Rehabilitation Act of 1973[39] and the Americans With Disabilities Act of 1990,[40] federal law now has a system of protections patterned after those set forth in Title VII creating a protected class for those possessing a physical or mental impairment that substantially limits one or more major life activities if such individuals are otherwise qualified to perform the essential functions of their job with the provision of reasonable accommodation for their disabilities by the employer. These protections, coupled with the provisions of Section 504 of the Rehabilitation Act governing students in educational settings receiving federal financial assistance and the protections of the Education of the Handicapped Act (now the Individuals with Disabilities Education Act),[41] suggest close scrutiny of mathematics assessment proposals to determine whether they present potential problems under these statutes. Each content standard needs to be scrutinized to determine whether the standard would serve as an unlawful bar to participation by a handicapped person in either an educational program or in employment. Of greatest importance for mathematics assessment will be the extent to which authentic tasks might present an unreasonable impediment to those with physically handicapping conditions or specific learning disabilities. The mechanisms for implementing both instruction and assessment of the standards will require similar scrutiny. Finally, assurance will be needed that employers do not unlawfully employ the standards to deny access to employment to any individuals with handicapping conditions. Here, the job-relatedness and business necessity requirements will be of the utmost importance and the burden of proof will rest on employers and, perhaps, the educators making certifications to employers.

Given these standards, the potential for disproportionate impact on the handicapped is high. Even if assessments rather than tests and subjective indicators such as portfolio assessments are used, legal risks are high given the potential impact of any mathematics assessment on decision-making for education or employment.

. . . .

GENDER-RELATED EFFECTS

Although there is potential for gender differences in almost any testing or assessment program, the issue may be particularly troublesome in mathematics assessment because of significant gender gaps in previous mathematics testing. Performance on such indicators as the Scholastic Aptitude Tests (now known as the Scholastic Achievement Test), the National Assessment for Educational Progress, and various vocational aptitude tests is consistently lower for females than males, particularly on higher-order tasks.

Allegations of gender bias in a mathematics assessment program could be subject to several types of legal challenge. Gender discrimination in education is directly addressed by the provisions of Title IX of the Education Amendments of 1972 [42] and its implementing regulations.[43] Title IX bars discrimination on the basis of sex in all educational programs and activities conducted by recipients of federal financial aid. Many states have similar provisions.[44] The legal analysis of Title IX challenges to gender disparities on mathematics assessments would probably follow the type of analysis used under Title VII of the Civil Rights Act of 1964[45] to assess discrimination in employment testing.[46] In addition, the provisions of Title VII barring gender discrimination in employment could also apply to use of the assessments in the workplace.

Judicial review of gender-related effects of assessment programs might also occur under the Equal Protection Clause of the Fourteenth Amendment to the U.S. Constitution or under analogous state constitutional provisions. In addition, approximately sixteen states have added equal rights amendments to their state constitutions in an effort to regulate gender discrimination. The state constitutional provisions differ to some extent in their interpretation and applicability by state courts.[47]

The use of mathematics assessments can, if challenged on the basis of gender discrimination, result in a judicial order to terminate the program, revalidate the assessment, create new assessments, or reconfigure the use of assessment results. It has also now been established that violations of Title IX can result in the award of monetary damages to provide recompense to the victims of unlawful gender discrimination.[48]

ACCESS TO INSTRUCTION FOR ALL STUDENTS

In addition to the legal issues that pertain to protected groups, there are legal challenges to educational reforms that can be mounted by any student. According to *Debra P.*, under the Due Process Clause of the Constitution, a program may be struck down by the courts if it is found to be "fundamentally unfair in that it may have covered matters not taught in the schools. . ."[49] Courts have deemed this a content validity issue.[50] Further, a test must be "a fair test of that which was taught" in order to withstand scrutiny under the Equal Protection Clause. Test fairness, under the *Debra P.* standard, hinges at least in part on test validity and reliability, two substantial technical hurdles that are apparently far from being resolved in the current discussions of the use of assessment to improve learning. Courts to date have not generally questioned the appropriateness of the content of what is taught except when challenges have been asserted on the basis of claims of denials of liberty,[51] establishment or free exercise of religion,[52] invasion of privacy, or free speech grounds,[53] which will be discussed briefly below.

The judicial holdings in *Debra P.* on curricular validity reinforced a behaviorist orientation in education at the time and fairly widespread attention began to be paid, for both educational and constitutional reasons, to requirements that teachers teach to the content of high-stakes tests. The constitutional standards set forth in *Debra P.*, and reiterated in subsequent federal cases,[54] will, for reasons to be discussed below, need to be considered in assessing the potential legal consequences of mathematics assessment, particularly if the individual stakes associated with assessment performance are high.

In 1983, the courts determined that Florida had met its burden of proving that these legal requirements were satisfied after the presentation of a massive set of surveys from local schools in which educators responded that they had addressed the competencies covered on the test. According to subsequent commentators, in part because of the test-curriculum match issue addressed by the courts, the national impact of the minimum competency movement was measurable; states were able to effectively overcome local control by requiring accountability and pass rates did increase, although in most instances school curriculum was diluted because the standards on the tests were set so low.[55] Because judicial scrutiny of governmental practices intensifies as the consequences of governmental action increase, the review of implementation efforts may be stricter than that adopted in the test-for-diploma programs if the consequences of assessment performance should include not only diploma denial, but also access to higher education or the workplace. When a scheme of national voluntary local participation in the initiative is created, legal responsibility for defending high stakes programs under the Due Process and Equal Protection Clauses of the Constitution, will rest with each participating state or local governmental entity.

Another issue worthy of further discussion is the legal consequences of moving from challenges to standardized practices, which was most often the case in the previous educational testing cases, to circumstances in which assessments involve open-ended tasks and more subjective judgments about the success of performance. Most prior test challenges were class actions brought against standardized testing practices. The use of different tasks or items that vary from school to school or state to state, with subjective assessments of performance, opens the door to thousands of potential individual cases. Courts can be expected to be reluctant to allow this expansion of litigation, particularly since it is on terrain that judges are ordinarily very hesitant to traverse. A plethora of individual discrimination cases will be difficult for members of protected groups to pursue. On the other hand, individual challenges by students and other groups on broad constitutional grounds may increase the number and rate of success as the more litigious members of our society apply their financial resources to efforts to obtain legal redress for educational grievances. This appears to be happening at present, for example, in cases involving challenges to disputed test scores from the Educational Testing Service.[56]

The Constitutional standards set forth in *Debra P.* v. *Turlington* require explicit recognition of the need to discuss state curriculum frameworks tied to standards.[57] Further, recognition that standards and assessments should be used not only to help measure progress, but also to implement progress, is another recognition of the constitutional due process standard embodied in *Debra P.* as well as the previous research literature indicating the extent to which a test can drive curriculum and instruction.

Finally, far more consideration needs to be given to how teachers can help to implement the educational reforms being proposed. Technical questions of assessment validity and reliability, governance, and policy implementation to one side, the initiative will work only if teachers can make it work. MSEB, as well as the National Council of Teachers of Mathematics, has already recognized that teachers will require considerable support to achieve this goal. Empowering teachers to meet the goals through teacher education and professional development will be critical to the success of the endeavor.

. . . .

ISSUES OF PERSONAL CHOICE

Additional constitutional issues may arise if programs using mathematics assessments follow the lead set by SCANS in its definitions of skills for the workplace. Some of these issues touch on some of the more controversial political matters presently confronting the nation. Privacy issues have rarely arisen in the past in debates over curriculum and assessment or testing. However, efforts to assess such variables as those enumerated by SCANS under the "personal qualities" and "interpersonal skills" categories may invite significant constitutional problems.

This nation, has a tradition of judicial protection of privacy interests in the face of government attempts to collect sensitive information about individuals or to use such information in order to make determinations about how government will treat individuals. For example, one federal court vetoed a junior high school's effort to use a drug use profile questionnaire to determine student placement in a drug abuse prevention program as a violation of the right to privacy.[58] There are also a series of issues regarding what government should do with potentially private information about a student once that information is acquired. For example, the Family Educational Rights and Privacy Act (FERPA) is designed to ensure parental and student access to a student's educational records and to place limitations on the release of information about a student without prior consent from the family.[59]   The statute was passed in an era in which Congress' privacy concerns were focused on such issues as student grade transcripts, letters of reference, or school psychologist's evaluation reports and the extent to which such information was being used outside a student's school to make potentially damaging judgments about the student. But the language and intent of FERPA apply to certification of mathematics proficiency and perhaps also to the information that may be used as the basis for granting certification, such as a student's performance on a particular assessment task. Given the interests of potential employers in obtaining access to such information, clear privacy protections must be in place.

Related to these privacy issues is a concern about the possibility of challenges on religious grounds to the content and assessment of standards. Policymakers should be prepared for the fact that some religious groups will have *bona fide* objections that standards or assessment techniques interfere with the free exercise of their religious rights or their free exercise of speech. The U.S. Supreme Court recognized exemptions from certain public education requirements for the Amish on the basis of this religious freedom argument.[60]   Such a challenge could be raised against some of the more fundamental and objective skills, such as higher-level mathematics or technology, as well as some of the more subjective assessment techniques. Assessments could conceivably be designed to identify a student's attitudes toward individual responsibility, sociability, or integrity that embodies religious or cultural biases. It is also possible that assessments could be implemented in a way that curtails an individual's opportunity to engage in free speech, such as might occur if the assessments of the SCANS "sociability" or "works with diversity" skills were applied to favor behavior that is in the currently popular terminology, "politically correct". The primary goal of the free speech clause of the First Amendment to the Constitution is, after all, the protection of all expressions of a point of view, even the most politically unpopular.[61]

EQUITY AND THE GOVERNANCE OF EDUCATION

With the proposal from some to establish a national assessment system that would truly be *national*, not *federal*, current reform initiatives acknowledge the long-standing tradition of state control of education. At the same time, however, the national reform movement encourages conformity in curriculum content, performance goals, and standards of assessment across states and localities. The Tenth Amendment to the U.S. Constitution requires that "the powers not delegated to the United States by the Constitution, nor prohibited by it to the states, are reserved to the states respectively, or to the people." Because education is not a power specifically given to the federal government, this doctrine of enumeration may be seen as barring efforts to create a federally mandated system of standards for states and localities. However, one power which is explicitly given the federal government is the power to regulate interstate and foreign commerce. A proposal to use assessment to reform the educational preparation of workers who will participate in interstate commerce may fall within the purview of the federal government's constitutional powers to regulate commerce. Given the breadth and depth of political enthusiasm for national education reform, the so-called "state's rights" concerns may be minimal, particularly in comparison to some of the other issues set forth in this paper. Further, there are any number of federal initiatives that have withstood scrutiny under the Tenth Amendment and have been vigorously enforced by the federal courts against the states under the terms of other constitutional provisions. For example, the Fourteenth Amendment's Due Process and Equal Protection Clauses have been used numerous times to enforce a national policy goal. In the education context, the most notable of these were the school desegregation cases, many of which inquired deeply into matters of local school curriculum and instruction.

The failure to adopt a federal system of curriculum standards and assessments presents the potential for fifty different sets of issues concerning validity, reliability, and fairness, with the ensuing possibility of fifty different sets of legal problems. If implementation is local, rather than at the state level, each local district could confront its own set of potential legal difficulties. These legal problems are accentuated whenever assessments are used for high-stakes decisions related to high school graduation, college admission, continuing education, and certification for employment. The technical problems inherent in a proposed system of high-stakes assessment are substantial. The potential legal difficulties and the enormity of the policy questions concerning such a proposal would urge great caution on the part of the proponents of such a system. A laudable goal such as that of the National Council on Education Standards and Testing (NCEST) to create a system of "tests worth teaching to" [62] can be lost among all of the other possible goals for the program, including: improving classroom instruction; improving learning outcomes for all students; informing students, parents, and teachers about student progress; measuring and holding students, schools, school districts, states, and the nation accountable for educational performance; assisting education policy-makers with programmatic decisions; certifying for future employment; credentials for college admission; etc.[63] The pursuit of multiple reform goals puts multiple pressures on both the psychometricians and preventive law specialists contemplating the manner in which the potential for challenges to such endeavors might be mounted. The possibility of having a separate set of practices in each governmental entity choosing to implement the program (not to mention any employer participating in the use of any resulting certification), creates the possibility for numerous legal challenges. Further, in any such legal challenge, the national or federal bodies with whom the localities are working might also face the risk of involvement. In short, it may be much wiser in the long run, particularly given that a truly national approach to these problems is being sought, to simply make this a *federal* effort and abandon the pretext of state and local control. Several commentators have already noted that all of the America 2000 initiatives are moving us inexorably toward a national curriculum.[64]

EQUITY AND ECONOMICS

Another set of potential legal issues centers around school finance and the current inequities from district to district and building to building in financial resources for education. Related to this is the potential impact of assessment information as a part of the inquiry in discussions of state takeovers of low-performing or educationally bankrupt school districts. In the past several years, state governments have become more willing to expand their exercise of responsibility for oversight of local education efforts. Some states, such as New Jersey, have implemented receiverships for certain low-performing districts. If mathematics assessment information or other educational accountability reports begin to inform state-level reviews of local district educational achievement, then such variables as the mathematics assessment will come to have very high stakes consequences not only for students but also for local school districts. As a result, such efforts might be subject to local district challenge on federal constitutional bases concerning due process and equal protection; they would, in addition, invite a broad array of state constitutional and statutory challenges concerning financing of education. Local district or individual school challenges to the use of assessment results might also be mounted under any state or federal school choice scheme in which assessment data could be used to limit a student's opportunity to attend a particular school.

Another practical issue, and one fraught with legal difficulties of another sort, is the disturbing question of how to pay for these ambitious initiatives. In his powerful reflection *Savage Inequalities*, Jonathan Kozol warns that the decision in *Brown v. Board of Education* "did not seem to have changed very much for children in the schools I saw, not, at least, outside of the Deep South"[65] and that "the dual society, at least in public education, seems in general to be unquestioned."[66] Further, most of the urban and less-affluent suburban schools he visited were untouched by school reform initiatives and, in the few instances where some reform initiatives had been tried, they amounted to little more than "moving around the same old furniture within the house of poverty. . . In public schooling, social policy has been turned back almost one hundred years."[67] At the core of all of these inequities, he finds, is a system of public finance of education which subsidizes and perpetuates these gross denials of educational opportunity.

The Implementation Task Force of the National Council for Education Standards and Testing suggests that equitable distribution of resources among districts and among schools within districts is a critical component for implementation at each level of government.[68] That group recognized that equity in funding is a key factor in the success of the endeavor[69] and will become a major issue in all of the states.[70]

Federal programs in the past have been critical in providing assistance for the educationally disadvantaged. Such endeavors will need to continue but should be linked tightly to the common content and performance standards.[71] NCEST in some respects seems to dismiss problems related to fiscal equity, hoping instead that national standards can create targets toward which educators can strive.[72] NCEST argues that states and local districts could work together to overcome deficiencies in resources.[73] Given the substantial difficulties that even one state, Texas, has had attempting to arrive at an equalization formula to address constitutional deficiencies with school funding, this seems an excessively optimistic position.[74]

Participants in the reform debate must maintain constant awareness of the possibility of unintended legal consequences. Once government defines minimum educational outcomes for all students and creates a presumption that sufficient educational services will be provided so that all students can meet this level of proficiency, then it may create an entitlement to an education that the federal courts have never previously been in a position to recognize for constitutional protection. In *San Antonio School District* v. *Rodriguez*,[75] the U.S. Supreme Court refused to recognize that education is a fundamental right under the Constitution; however, if a fundamental right is, in essence, created as the result of the creation of an entitlement, then the level of judicial scrutiny of a governmental practice may be subject to the burdensome "strict scrutiny" level of analysis of practices that work to deny citizens' fundamental interests, a burden nearly impossible for government to meet.

A related issue concerns the fact that the government will have created a legitimate expectation on the part of students that school attendance will result in attainment of a certain level of mathematics skills. This also creates a need to assess whether the doors previously closed by state court judges to claims of "educational malpractice" may be wedged open again as a result of the new national standards and goals.[76]

William Clune identifies four generic problems confronting efforts to enhance student achievement: poor understanding of effective practice (weak technology, that is, lack of understanding of which practices produce improved learning); serious problems of policy implementation (central control can do little to affect the activities of millions of teachers and learners across the nation); serious problems of political organization and policy formation (effective educational policy must be carefully designed and tightly coordinated); and significant cost constraints (massive infusions of new capitol would be needed to subsidize major change).[77] One cannot hope, Clune asserts, to successfully pursue strong educational goals through the use of weak policy instruments; he views efforts at reform through the use of educational indicators and assessments as requiring, in particular, further development of assessments that are technically defensible and efforts to influence instructional content and practice as requiring tighter systems to guide instruction, perhaps including teacher education and systems to maximize teacher participation, enthusiasm, and responsibility, and greater focus on curriculum to promote higher-order thinking and problem solving. Each of these concerns has an analog in the legal issues discussed above. Without a satisfactory solution to each of these problems, the legal consequences could be substantial. In particular, specific attention must be paid to the impact of these proposals on educationally disadvantaged students. From a policy perspective, issues of equity should be of the utmost importance. From a legal perspective, it may be those who have traditionally been the most educationally disadvantaged who will be able to bring the most successful legal challenges to the endeavor. From an economic perspective, a failure to effectively address the needs of *all* students will have devastating consequences for the future economic welfare of the entire nation.

CONCLUSION

This paper provides a brief summary of the principal legal and policy issues that might arise in challenges to a mathematics assessment initiative by members of protected groups traditionally underserved by the nation's schools, by any student who performs poorly on an assessment, or by individual school districts. Enhanced educational attainment in mathematics is a goal with which few could disagree. However, educators and public policymakers should take care that all schools are provided sufficient resources to allow them to effectively meet that goal and that all students, no matter their race, ethnicity, language, background, or handicapping condition are given a fair opportunity to learn and a fair opportunity to demonstrate their learning through assessments. Finally, without an adequate system of financing mathematics education and assessment in all schools, no effort at education reform will succeed.

ENDNOTES

1 This paper was considerably influenced by a previous paper by the author commissioned by the Secretary's Commission on Achieving Necessary Skills of the U.S. Department of Labor.

2 *Brown v. Board of Education*, 347 U.S. 483 (1954).

3 *Debra P. v. Turlington*, 474 F. Supp. 244 (M.D. Fla. 1979); aff'd in part, rev'd in part, 644 F. 2d 397 (5th Cir. 1981); reh. en banc den.

4 474 F. Supp. at 247.

5 Id. at 249.

6 42 U.S.C. 2000d.

7 474 F. Supp. at 252.

8 Id. citing *Armstead v. Starkville Municipal Separate School District*, 461 F. 2d 276 (5th Cir. 1972).

9 644 F. 2d 397, at 400.

10 Id. at 402.

11 See text accompanying endnotes 2-6.

12 644 F. 2d 397, 406-407.

13 644 F. 2d at 407.

14 Although the latter is not a basis for a legal claim in most circumstances, it correlates with race and ethnicity and may thus result in a basis for a legal challenge.

15 An item from the February 1991 Maryland School Performance Assessment Program Grade 8 Mathematics Assessment teacher's guide involves a task asking students to develop a survey plan to collect information on potential respondents to assist a developer's efforts to build a new restaurant. The lowest-scoring sample student answer is "I would ask people in the rich part of the county." Without doubt, that response lacks a richness of detail that reflects much understanding of sampling methodology even at the eighth grade level, but for a low-income student who could never contemplate having the opportunity to be a developer, the sample answer says it all.

16 508 F. 2d 1017 (5th Cir. 1975).

17 See also, *Hobson v. Hansen*, 269 F. Supp. 401 (D.D.C 1967), aff'd sub nom *Smuck v. Hansen*, 408 F. 2d 175 (D.C. Cir. 1969) (en banc).

18 *Larry P. v. Riles*, 343 F. Supp. 1306 (N.D. Cal. 1972), aff'd 502 F. 2d 963 (9th Cir. 1974).

19 There are also issues concerning both education and employment of persons with limited English proficiency (LEP); these issues are not addressed here on the assumption (perhaps erroneous) that courts will find English proficiency requirements quite acceptable for the nation's future workplaces. However, even if this assumption is true, there is another set of legal issues, unaddressed here, concerning the rights of LEP students to education that meets their special needs.

[20] *What Work Requires of Schools: A SCANS Report for America 2000.* The Labor Secretary's Commission on Achieving Necessary Skills, U.S. Department of Labor (hereinafter SCANS), 1991, p. 24.

[21] SCANS, p. 25.

[22] SCANS, pp. 26-27.

[23] SCANS, p. 27.

[24] 42 U.S.C. 2000e - 2(a)(1).

[25] 42 U.S.C. 2000e - 2(h).

[26] *Griggs* v. *Duke Power Co.,* 401 U.S. 424 at 431 (1971).

[27] See P. Patterson, "Employment Testing and Title VII of the Civil Rights Act of 1964" in Gifford and O'Connor, pp. 93-95.

[28] *Albemarle Paper Co.* v. *Moody,* 422 U.S. 405 at 425 (1975).

[29] *Griggs* and *Albemarle.*

[30] *Albemarle,* at 431.

[31] *Albemarle,* op. cit.

[32] See B. Schlei and P. Grossman, *Employment Discrimination Law* (1983), pp. 98-161 and 1985 Supp. p. 18; See *Test Policy and the Politics of Opportunity Allocation: The Workplace and the Law,* B. Gifford, ed, Klover, Boston (1989).

[33] 422 U.S. at 434.

[34] 422 U.S. at 435.

[35] Note also that to the extent that proposals may be implemented in a manner not analogous to a scored test, but rather as a less uniform assessment according to subjective criteria, Title VII is applicable and the *Griggs* standard is followed. See Schlei, B. L. & Grossman, P. (1983). *Employment Discrimination Law* (2nd ed.). Washington, DC: Bureau of National Affairs, Inc. pp. 162-190, and 1993-84 *Cum-Supp.,* pp. 21-23.

[36] P.L. 102-166.

[37] Senate sponsors of the law, including Senators Danforth, Kennedy, and Dole, and the administration created a specific legislative history for the law stating that the terms "business necessity" and "job-relatedness" are intended to reflect the concepts enunciated by the Supreme Court in *Griggs* v. *Duke Power Co.,* 401 U.S. 424 (1971), and in the other Supreme Court decisions prior to *Wards Cove Packing Co.* v. *Atonio,* 490 U.S. 642 (1989). When a decision-making process includes particular, functionally integrated practices that are components of the same criterion, standard, method of administration, or test, such as the height and weight requirements designed to measure strength in *Dothard* v. *Rawlinson,* 433 U.S. 321 (1971), the particular, functionally integrated practices may be analyzed as one employment practice."

[38] P.L. 102-166, Sec. 106.

[39] 29 U.S.C. 701 et. seq.

[40] 42 U.S.C. 12101 et. seq.

[41] 20 U.S.C. 1400 et. seq.

[42] 20 U.S.C. 1681-1687, as amended by the Civil Rights Restoration Act of 1987, codified at 20 U.S.C. 1687.

[43] 34 C.F.R. 86.1-86.70.

[44] See, for example, Massachusetts General Laws Ann. ch. 76, sec. 5.

[45] 42 U.S.C. 2000e.

[46] See K. Connor and E. Vargyas, "The Legal Implications of Gender Bias in Standardized Testing," *Berkeley Women's Law Journal*, (1992), pp. 13-89 for an excellent analysis of gender discrimination law as it applies to testing.

[47] Id.

[48] Franklin v. Gwinnett County Public Schools, 112 S. Ct. 1028 (1992).

[49] 644 F. 2d at 403, emphasis in original.

[50] 644 F. 2d 397, 404.

[51] See *Meyer v. Nebraska*, 262 U.S. 390 (1923).

[52] See *Wisconsin v. Yoder*, 406 U.S. 205 (1972).

[53] See *West Virginia State Board of Education v. Barnette*, 319 U.S. 624 (1943).

[54] *Brookhart v. Peoria*, 697 F. 2d 182 (7th Cir. 1982). See *Anderson v. Banks*, 520 F. Supp. 472 (S.D. Ga., 1981), appeal from subsequent order dismissed sub nom. *Johnson v. Sikes*, 730 F. 2d 644 (11th Cir. 1984).

[55] E. Baker and R. Stites, "Trends in Testing in the USA" in *The Politics of Curriculum and Testing*, S.H. Fuhrman and B. Malen (eds.) (1991), pp. 148-149.

[56] "Court Orders Testing Service to Release Disputed Scores": *The Chronicle of Higher Education*, September 2, 1992.

[57] *Raising Standards for American Education: A Report to Congress, the Secretary of Education, the National Education Goals Panel, and the American People.* The National Council on Education Standards and Testing (hereinafter NCEST), Washington D.C. 1992, p. 7.

[58] *Merriken v. Cressman*, 364 F. Supp. 913 (E.D. Pa. 1973).

[59] 20 U.S.C. 1232g et. seq.; 34 C.F.R. Part 99.

[60] *Wisconsin v. Yoder*, 406 U.S. 205 (1971).

[61] L. Tribe, *American Constitutional Law* (2nd ed.) (1988), Mineola, NY: The Foundation Press, Inc. pp. 785-1061; M. Yudof, D. Kirp, T. VanGeel, and B. Levin, *Educational Policy and the Law* (2nd ed.), (1982), Berkeley, CA: McCutchan Publishing. pp. 205-212.

[62] NCEST, p. 6.

[63] NCEST, pp. 5, 6.

[64] E. Baker and R. Stites (1991). Trends in testing in the USA. *Politics of Education Association yearbook 1990.* (p. 152) London: Taylor & Francis.

[65] J. Kozol, *Savage Inequalities: Children in America's Schools.* New York, Crown Publishers, Inc. (1991), p. 3.

[66] Id. p. 4.

[67] Id.

[68] NCEST Implementation Task Force, p. G-7.

[69] Id. p. G-13.

[70] Id.

[71] NCEST Standards Task Force Report, p. E-13.

[72] Id. p. E-15.

[73] Id.

[74] Lonnie Harp, "Texas Finance Bill Signed Into Law, Challenges Anticipated," *Education Week,* 9 June 1993; Lonnie Harp, "Impact of Texas Finance Law, Budget Increase Gauged," *Education Week,* 16 June 1993; Millicent Lawton, "Alabama Judge Sets October Deadline for Reform Remedy," 23 June 1993.

[75] 411 U.S. 1 (1972).

[76] See e.g., E.T. Connors, *Educational Tort Liability and Malpractice,* 1981, pp. 148-158.

[77] W. Clune, "Educational policy in a situation of uncertainty; or, how to put eggs in different baskets," in Fuhrman and Malen, op. cit., pp. 132-133.

STUDY GROUP ON GUIDELINES FOR MATHEMATICS ASSESSMENT

Jeremy Kilpatrick, Chair
Regents Professor of Mathematics
Education
University of Georgia
Athens, GA

Janice Arceneaux
Magnet Specialist
Houston Independent School
District
Houston, TX

Lloyd Bond
Professor of Educational Research
University of North Carolina
Greensboro, NC

Felix Browder
Professor of Mathematics
Rutgers University
New Brunswick, NJ

Philip C. Curtis, Jr.
Professor of Mathematics
University of California
at Los Angeles
Los Angeles, CA

Jane D. Gawronski
Superintendent
Escondido Union High School
District
Escondido, CA

Robert L. Linn
Professor of Education
University of Colorado
Boulder, CO and
Co-Director, Center for Research
on Evaluation, Standards, and
Student Testing at UCLA

Sue Ann McGraw
Mathematics Teacher
Lake Oswego High School
Lake Oswego, OR

Robert J. Mislevy
Principal Research Scientist
Educational Testing Service
Princeton, NJ

Alice Morgan-Brown
Statewide Director for Academic
Champions of Excellence Program
Morgan State University
Baltimore, MD

Andrew Porter
Director, Wisconsin Center for
Education Research
Professor of Education Psychology
University of Wisconsin
Madison, WI

Marilyn Rindfuss
National Mathematics Consultant
The Psychological Corporation
San Antonio, TX

Edward Roeber
Director, Student Assessment
Programs
Council of Chief State School
Officers
Washington, DC

Maria Santos
Mathematics & Science Supervisor
San Francisco Unified School
District
San Francisco, CA

Cathy Seeley
Director of Pre-college Programs
Charles A. Dana Center for
Mathematics and Science Education
University of Texas
Austin, TX

Edward A. Silver
Senior Scientist, Learning Research
& Development Center
Professor of Education
University of Pittsburgh
Pittsburgh, PA

MSEB Members

* Executive Committee

Hyman Bass, Chair * (1996)
Professor of Mathematics
Columbia University
New York, NY

Mary M. Lindquist,
Vice Chair * (1994)
Callaway Professor of Mathematics
Education
Columbus College
Columbus, GA

Maria A. Berriozabal (1993)
Former City Councilwoman
San Antonio, TX

Sylvia T. Bozeman (1995)
Chair, Department of Mathematics
Spelman College
Atlanta, GA

Sadie C. Bragg (1996)
Acting Dean of Academic Affairs
Borough of Manhattan Community
College
The City University of New York
New York, NY

Diane J. Briars (1995)
Director, Division of Mathematics
Pittsburgh Public Schools
Pittsburgh, PA

Patricia Chavez (1996)
Statewide Executive Director
New Mexico Mathematics,
Engineering, Science Achievement
Albuquerque, NM

Loring Coes, III * (1993)
Chair, Department of Mathematics
Rocky Hill School
East Greenwich, RI

Nancy S. Cole * (1994)
Executive Vice President
Educational Testing Service
Princeton, NJ

Rita R. Colwell (1993)
President,
Maryland Biotechnology Institute
University of Maryland
College Park, MD

Shari Anne Wilson Coston
(1996)
Executive Director, Arkansas
Education Renewal Consortium
Conway, AK

David Crippens (1993)
Senior Vice President
Educational Enterprises
KCET-TV, PBS
Los Angeles, CA

Gilbert J. Cuevas * (1995)
Professor of Education
University of Miami
Coral Gables, FL

Philip A. Daro * (1993)
Executive Director
University of California
California Mathematics Project
Oakland, CA

Daniel T. Dolan (1993)
Associate Director, PIMMS Project
Wesleyan University
Middletown, CT

Sue T. Dolezal (1995)
Mathematics Instructor
Sentinel High School
Missoula, MT

Dale Ewen (1994)
Vice President
Academic and Student Services
Parkland College
Champaign, IL

Joseph A. Fernandez (1994)
Chancellor
New York City Board of Education
Brooklyn, NY

Rol Fessenden (1996)
Director of Inventory Control
L.L. Bean, Inc.
Freeport, ME

Gary Froelich (1994)
Secondary Schools Project Manager
Consortium for Mathematics and Its
Applications
Lexington, MA

John Gage (1994)
Director, Science Office
Sun Microsystems
Mountain View, CA

Ramesh I. Gangolli (1994)
Professor of Mathematics
University of Washington
Seattle, WA

Linda M. Gojak * (1995)
Chair (K–8) Mathematics
Department
Hawken School
Lyndhurst, OH

Jackie Goldberg (1993)
City Councilwoman
Los Angeles, CA

Jacqueline E. Goodloe * (1995)
Elementary Mathematics Resource
Teacher
Burrville Elementary School,
Washington, DC

Ronald L. Graham (1995)
Adjunct Director
Information Sciences Division
AT&T Bell Laboratories
Murray Hill, NJ

Norbert Hill (1993)
Executive Director, American Indian
Science and Engineering Society
Boulder, CO

Patricia S. Henry (1994)
President, The National PTA
Lawton, OK

Elizabeth M. Jones (1995)
K–12 Mathematics Consultant
Mathematics Office
Lansing School District
Lansing, MI

James S. Kahn * (1994)
President and CEO
Museum of Science and Industry
Chicago, IL

Harvey B. Keynes (1993)
Professor of Mathematics and
Director of Special Projects
University of Minnesota
Minneapolis, MN

Glenda T. Lappan * (1996)
Professor of Mathematics
Michigan State University
East Lansing, MI

Paul G. LeMahieu (1996)
Director, Delaware Education
Research & Development Center
University of Delaware
Newark, DE

Sue Ann McGraw (1994)
Mathematics Instructor
Lake Oswego High School
Lake Oswego, OR

Ruth R. McMullin * (1993)
Acting President & CEO
Publishing Group
Harvard Business School
Boston, MA

Fernando Oaxaca (1994)
President, Coronado
Communications
Los Angeles, CA

Jack Stanley Price (1996)
Co-Director, Center for Science
and Mathematics Education
California State Polytechnic
University
Pomona, CA

C.R. Richmond (1996)
Director, Science Education
Programs and External Relations
Oak Ridge National Laboratory
Oak Ridge, TN

William E. Spooner (1996)
Chief Consultant, Science Division
North Carolina Department of
Public Instruction
Raleigh, NC

Lee V. Stiff (1996)
Associate Professor of Mathematics
North Carolina State University
Raleigh, NC

Robert A. Strickland (1996)
Senior Consultant and Director for
Information Technology
Harvard Business School
Boston, MA

Stephanie Sullivan (1996)
Executive Director
Rhode Island Mathematical Sciences
Education Coalition
Providence, RI

Richard A. Tapia (1995)
Professor of Mathematical Sciences
Department of Computational and
Applied Mathematics
Rice University
Houston, TX

Daniel J. Teague (1994)
Mathematics Instructor
North Carolina School of
Mathematics and Science
Durham, NC

John S. Toll, (1993)
President, Universities Research
Association, Inc.
Washington, DC, and Chancellor
Emeritus and Professor of Physics
University of Maryland
College Park, MD

Philip Uri Treisman (1993)
Professor of Mathematics
University of Texas at Austin
Austin, TX

Alvin W. Trivelpiece *
(Chair 1990–1993)
Director, Oak Ridge National
Laboratory
Oak Ridge, TN

Larry DeVan Williams (1993)
Mathematics Instructor
Eastwood Middle School
Tuscaloosa, AL

James J. Wynne (1993)
Manager, Biology and Molecular
Science
IBM T.J. Watson Research Center
Yorktown Heights, NY